Physical Separation and Enrichment

Physical Separation and Enrichment

Special Issue Editor
Saeed Farrokhpay

MDPI • Basel • Beijing • Wuhan • Barcelona • Belgrade

Special Issue Editor
Saeed Farrokhpay
Université de Lorraine
France

Editorial Office
MDPI
St. Alban-Anlage 66
4052 Basel, Switzerland

This is a reprint of articles from the Special Issue published online in the open access journal *Minerals* (ISSN 2075-163X) from 2018 to 2020 (available at: https://www.mdpi.com/journal/minerals/special_issues/Physical_Separation_Enrichment).

For citation purposes, cite each article independently as indicated on the article page online and as indicated below:

LastName, A.A.; LastName, B.B.; LastName, C.C. Article Title. *Journal Name* **Year**, *Article Number*, Page Range.

ISBN 978-3-03928-436-8 (Pbk)
ISBN 978-3-03928-437-5 (PDF)

Contents

About the Special Issue Editor . . . . **vii**

Preface to "Physical Separation and Enrichment" . . . . **ix**

Saeed Farrokhpay
Editorial for the Special Issue: "Physical Separation and Enrichment"
Reprinted from: *Minerals* **2020**, *10*, 173, doi:10.3390/min10020173 . . . . **1**

Sunil Kumar Tripathy, Y Rama Murthy, Veerendra Singh, Saeed Farrokhpay and Lev O. Filippov
Improving the Quality of Ferruginous Chromite Concentrates Via Physical Separation Methods
Reprinted from: *Minerals* **2019**, *9*, 667, doi:10.3390/min9110667 . . . . **4**

Jianwu Zeng, Xiong Tong, Fan Yi and Luzheng Chen
Selective Capture of Magnetic Wires to Particles in High Gradient Magnetic Separation
Reprinted from: *Minerals* **2019**, *9*, 509, doi:10.3390/min10020173 . . . . **23**

Kwanho Kim and Soobok Jeong
Separation of Monazite from Placer Deposit by Magnetic Separation
Reprinted from: *Minerals* **2019**, *9*, 149, doi:10.3390/min9030149 . . . . **35**

Yuekan Zhang, Peikun Liu, Lanyue Jiang, Xinghua Yang and Junru Yang
Numerical Simulation of Flow Field Characteristics and Separation Performance Test of Multi-Product Hydrocyclone
Reprinted from: *Minerals* **2019**, *9*, 300, doi:10.3390/min9050300 . . . . **47**

Yuekan Zhang, Peikun Liu, Lanyue Jiang and Xinghua Yang
The Study on Numerical Simulation and Experiments of Four Product Hydrocyclone with Double Vortex Finders
Reprinted from: *Minerals* **2019**, *9*, 23, doi:10.3390/min9010023 . . . . **61**

Lanyue Jiang, Peikun Liu, Yuekan Zhang, Xinghua Yang and Hui Wang
The Effect of Inlet Velocity on the Separation Performance of a Two-Stage Hydrocyclone
Reprinted from: *Minerals* **2019**, *9*, 209, doi:10.3390/min9040209 . . . . **78**

Ali Davoodi, Gauti Asbjörnsson, Erik Hulthén and Magnus Evertsson
Application of the Discrete Element Method to Study the Effects of Stream Characteristics on Screening Performance
Reprinted from: *Minerals* **2019**, *9*, 788, doi:10.3390/min9120788 . . . . **97**

Changshuai Chen, Longlong Wang, Zhenfu Luo, Yuemin Zhao, Bo Lv, Yanhong Fu and Xuan Xu
The Effect of the Characteristics of the Partition Plate Unit on the Separating Process of −6 mm Fine Coal in the Compound Dry Separator
Reprinted from: *Minerals* **2019**, *9*, 215, doi:10.3390/min9040215 . . . . **110**

Xin Du, Chao Liang, Donglai Hou, Zhiming Sun and Shuilin Zheng
Scrubbing and Inhibiting Coagulation Effect on the Purification of Natural Powder Quartz
Reprinted from: *Minerals* **2019**, *9*, 140, doi:10.3390/min9030140 . . . . **128**

Lu Yang, Zhenna Zhu, Xin Qi, Xiaokang Yan and Haijun Zhang
The Process of the Intensification of Coal Fly Ash Flotation Using a Stirred Tank
Reprinted from: *Minerals* **2018**, *8*, 597, doi:10.3390/min8120597 . . . . **142**

Caroline C. Gonçalves and Paulo F. A. Braga
Heavy Mineral Sands in Brazil: Deposits, Characteristics, and Extraction Potential of
Selected Areas
Reprinted from: *Minerals* **2019**, *9*, 176, doi:10.3390/min9030176 . . . . **156**

**Xu Wang, Wenqing Qin, Fen Jiao, Congren Yang, Yanfang Cui, Wei Li, Zhengquan Zhang
and Hao Song**
Mineralogy and Pretreatment of a Refractory Gold Deposit in Zambia
Reprinted from: *Minerals* **2019**, *9*, 406, doi:10.3390/min9070406 . . . . **171**

About the Special Issue Editor

Saeed Farrokhpay obtained his Ph.D. in Mineral and Material Engineering from Ian Wark Research Institute, University of South Australia in 2005. He is a research scientist at GeoRessources, University of Lorraine in France, where he has worked for the past five years. Previously, he worked as a researcher at JKMRC, University of Queensland and University of South Australia. His research interests involve mineral separation and processing complex ores. He published more than 80 papers in highly ranked peer reviewed journals and conference proceedings.

Preface to "Physical Separation and Enrichment"

This Special Issue includes 12 papers from around the world on topics related to physical separation and enrichment in mineral processing. Physical separation has been used in mineral industry for centuries to separate valuable minerals from gangues using differences in their physical properties. Physical separation methods have several advantages over other mineral processing techniques due to their high efficiency, low capital and operating costs, no additional chemicals required, and consequently, lower environmental hazard. They can be applied to the ores from mines or tailings, or at the recycling stage for scavenging the desired elements. Physical separation methods are also used to upgrade the valuable minerals before the main beneficiation process when the valuable minerals are at low grades, or in purification of flotation products in processing complex ores. In this Special Issue, gravity and magnetic separation, hydrocyclones, screens, dry separators, scrubbing, and application of physical separation in processing complex orebodies are discussed.

Saeed Farrokhpay
Special Issue Editor

Editorial

Editorial for the Special Issue: "Physical Separation and Enrichment"

Saeed Farrokhpay

GeoRessources, University of Lorraine, 54000 Nancy, France; saeed.farrokhpay@univ-lorraine.fr

Received: 4 February 2020; Accepted: 8 February 2020; Published: 14 February 2020

1. Introduction

Physical separation methods have been used in mineral industry for centuries to separate valuable minerals from gangues using differences in their physical properties. They have several advantages over other mineral processing techniques due to their high efficiency, low capital and operating costs, no addition of chemicals and consequently, less environmental hazard. They can be applied to ores from mines and tailings or at the recycling stage for scavenging the desired elements. They are also used to upgrade valuable minerals (i.e., pre-concentration) before the main beneficiation process when such minerals are at low grades and their separation would be costly [1], or in purification of flotation products in processing complex ores [2,3].

2. The Special Issue

This Special Issue includes twelve papers from around the world on topics related to "Physical Separation and Enrichment" in mineral processing. These papers discuss the latest findings on using physical separation methods in mineral processing. They include papers discussing gravity and magnetic separation, hydrocyclones, screens, dry separators, scrubbing, and application of physical separation in processing complex orebodies.

Gravity concentration is one of the oldest industrial methods which is widely used for mineral separation. Its importance has not decreased in the 21st century even with the invention of froth flotation. In this method, gravity, or centrifugal force, is used to separate mixed particles either in suspension or in dry form. A comprehensive review of gravity separation was recently published [4]. Magnetic separation also uses the magnetic properties of different minerals, and the most common application of this method, as one may expect, is separation of iron and iron-bearing minerals [5]. In this Special Issue, Tripathy et al. [6] discuss improving the quality of ferruginous chromite concentrates using gravity and magnetic separation. The value of chromite ores is determined by their iron and chromium contents. Therefore, reprocessing chromite concentrates by physical separation to enhance the chromium-to-iron ratio is an important issue which is covered in this paper. Zeng et al. [7] also discuss selective capture of magnetic wires in high gradient magnetic separation which achieves an effective separation of fine weakly magnetic minerals using numerous small magnetic wires. In addition, Kim and Jeong [8] also used a magnetic method to separate rare earth elements in a deposit in North Korea. Rare earth elements have vital applications in modern technology, and their importance is continuously growing in the world.

Hydrocyclones are devices that can effectively separate multi-phase mixtures of particles with different densities or sizes based on centrifugal sedimentation principles. Traditional hydrocyclones can only generate two products with different size fractions after one classification which may not meet the requirements for the narrow size fractions in fine particle classification. Therefore, Zhang et al. [9,10] discuss multi-product hydrocyclones via numerical simulation of flow field characteristics and separation performance tests of multi-products, as well as numerical separation and experiments of four product hydrocyclones. Their simulation results showed that, in contrast with the traditional single overflow pipe, there are two turns in the internal axial velocity direction of the hydrocyclone with the double overflow pipe structure. Zhang et al. [9,10] provide evidence for understanding the flow field distribution in hydrocyclones and the development of multi-product grading instruments in terms of both theory and industrial applications. Jiang et al. [11] also discuss the effect of inlet velocity on the separation performance of a two-stage hydrocyclone. The entrainment of coarse particles in overflow, and fine particles in underflow, are two inevitable phenomena in the hydrocyclone separation process which can result in a wide product size distribution. Hence, this study proposed a two-stage hydrocyclone, and the effects of the inlet velocity on the hydrocyclone were investigated using computational fluid dynamics.

Screening is a key operation in a crushing plant which ensures adequate product quality of aggregates in mineral processing. It is affected by the relative difference among various properties such as particle shape, size distribution and material density. Davoodi et al. [12] discuss the application of the discrete element method to study the effects of stream characteristics on the screening performance. Discrete element method is a method for analysing the interactions among individual particles and between particles and a screen deck in a controlled environment. This paper demonstrates that denser particles have a higher probability of passage due to their higher stratification rate which increases the probability of contacting a particle with the screen deck.

The effect of the partition plate unit on the separating of fine coal particles in a compound dry separator is reported by Chen et al. [13]. Compound dry separation technology is often applied for the separation of coal of size fractions above 6 mm, and it has not been widely used in finer fractions. In this paper, the effect of partition plate unit's characteristics on both average density of particles in the bed uniformly and the final separation results for fine coal particles are discussed. The results show that characteristics of the partition plate unit had important effects on the separating process of fine coal particles in a compound dry separator.

Du et al. [14] also contributed to this Special Issue by discussing scrubbing and the inhibiting coagulation effect on the purification of quartz particles. The low removal efficiency of fine clay impurities in natural quartz is the main problem affecting the practical usage of this resource. In this paper, a combined physical purification process, including sieving, scrubbing and centrifugation, is discussed to remove the clay impurities. In another paper, Yang et al. [15] discuss the process of the intensification of coal fly ash using a stirred tank.

Application of physical separation methods to process heavy mineral sand in Brazil is discussed by Gonçalves and Braga [16]. They obtained minimum 70% recovery of the desired minerals by using physical separation units (such as a shaking table and a magnetic separator). Wang et al. [17] also studied a gold deposit located in Zambias where most of the gold particles were less than 10 µm. They obtained more than 90% recovery by using a gravity–flotation combined beneficiation process to recover the liberated coarse and fine gold particles. Gravity–flotation combined beneficiation pre-treatment, in fact, provided a feasible method for the complex refractory gold ore.

3. Summary

This Special Issue is a good example of the growing number of technical and scientific activities around the world trying to develop and understand various methods to separate valuable minerals. We hope this issue will shed light on various aspects of physical separation and enrichment and enhance the knowledge and scientific debate in this field of engineering research.

Acknowledgments: The Guest Editor would like to thank all authors, reviewers, the editor Mr. Irwin Liang and the editorial staff of the *Minerals* journal for their timely efforts to successfully complete this Special Issue.

References

1.   Farrokhpay, S.; Fornasiero, D.; Filippov, L. Preconcentration of nickel in laterite ores using physical separation methods. *Miner. Eng.* **2019**, *141*, 105892. [CrossRef]

2. Yue, T.; Han, H.; Hu, Y.; Wei, Z.; Wang, J.; Wang, L.; Sun, W.; Yang, Y.; Sun, L.; Liu, R.; et al. Beneficiation and purification of tungsten and cassiterite minerals using Pb-BHA complexes flotation and centrifugal separation. *Minerals* **2018**, *8*, 566. [CrossRef]
3. Foucaud, Y.; Dehaine, Q.; Filippov, L.; Filippova, I. Application of falcon centrifuge as a cleaner alternative for complex tungsten ore processing. *Minerals* **2019**, *97*, 448. [CrossRef]
4. Das, A.; Sarkar, B. Advanced gravity concentration of fine particles: A review. *Miner. Process. Extr. Metall. Rev.* **2018**, *39*, 359. [CrossRef]
5. Rao, S.R.R. *Resource Recovery and Recycling from Metallurgical Wastes*; Elsevier: Amsterdam, The Netherlands, 2006.
6. Tripathy, S.K.; Rama, Y.; Murthy, V.S.; Farrokhpay, S.; Filippov, L. Improving the quality of ferruginous chromite concentrates via physical separation methods. *Minerals* **2019**, *9*, 667. [CrossRef]
7. Zeng, J.; Tong, X.; Yi, Y.; Chen, L. Selective capture of magnetic wires to particles in high gradient magnetic separation. *Minerals* **2019**, *9*, 509. [CrossRef]
8. Kim, K.; Jeong, S. Separation of monazite from placer deposit by magnetic separation. *Minerals* **2019**, *9*, 149. [CrossRef]
9. Zhang, Y.; Liu, P.; Jiang, L.; Yang, X.; Yang, L. Numerical simulation of flow field characteristics and separation performance test of multi-product hydrocyclone. *Minerals* **2019**, *9*, 300. [CrossRef]
10. Zhang, Y.; Liu, P.; Jiang, L.; Yang, H. The study on numerical simulation and experiments of four product hydrocyclone with double vortex finders. *Minerals* **2019**, *9*, 23. [CrossRef]
11. Jiang, L.; Liu, P.; Zhang, Y.; Yang, X.; Wang, H. The effect of inlet velocity on the separation performance of a two-stage hydrocyclone. *Minerals* **2019**, *9*, 209. [CrossRef]
12. Davoodi, A.; Asbjörnsson, G.; Hulthén, E.; Evertsson, M. Application of the discrete element method to study the effects of stream characteristics on screening performance. *Minerals* **2019**, *9*, 788. [CrossRef]
13. Chen, C.; Wang, L.; Luo, Z.; Zhao, Y.; Lv, B.; Fu, Y.; Xu, X. The Effect of the characteristics of the partition plate unit on the separating process of −6 mm fine coal in the compound dry separator. *Minerals* **2019**, *9*, 215. [CrossRef]
14. Du, X.; Liang, C.; Hou, D.; Sun, Z.; Zheng, S. Scrubbing and inhibiting coagulation effect on the purification of natural powder quartz. *Minerals* **2019**, *9*, 140. [CrossRef]
15. Yang, L.; Zhu, Z.; Qi, X.; Yan, X.; Zhang, H. The process of the intensification of coal fly ash flotation using a stirred tank. *Minerals* **2018**, *8*, 597. [CrossRef]
16. Gonçalves, C.C.; Braga, P.F.A. Heavy mineral sands in Brazil: Deposits, characteristics, and extraction potential of selected areas. *Minerals* **2019**, *9*, 176. [CrossRef]
17. Wang, X.; Qin, W.; Jiao, F.; Yang, C.; Cui, Y.; Li, W.; Zhang, Z.; Song, H. Mineralogy and pre-treatment of a refractory gold deposit in Zambia. *Minerals* **2019**, *9*, 406. [CrossRef]

Article

Improving the Quality of Ferruginous Chromite Concentrates Via Physical Separation Methods

Sunil Kumar Tripathy [1,2,*], Y Rama Murthy [1], Veerendra Singh [1], Saeed Farrokhpay [2] and Lev O. Filippov [2]

[1] Research and Development Division, Tata Steel Ltd., Jamshedpur 831001, India; yrama.murty@tatasteel.com (Y.R.M.); veerendra.singh@tatasteel.com (V.S.)
[2] GeoRessources, University of Lorraine, CNRS, UMR 7359, F54505 Nancy, France; saeed.farrokhpay@univ-lorraine.fr (S.F.); lev.filippov@univ-lorraine.fr (L.O.F.)
* Correspondence: sunil-kumar.tripathy@univ-lorraine.fr or sunilkr.tripathy@gmail.com; Tel.: +33-372-744-547 or +33-625-329-139

Received: 8 October 2019; Accepted: 28 October 2019; Published: 29 October 2019

Abstract: The low chromium-to-iron ratio of chromite ores is an important issue in some chromite deposits. The value of the chromite ore is indeed dictated in the market by its iron, as well as its chromium content. In the present study, a chromite concentrate was reprocessed by gravity (spiral concentrator) and magnetic separation to enhance the chromium-to-iron ratio. Also, detailed characterization studies including automated mineralogy were carried out to better understand the nature of the samples. Enhancing the chromium-to-iron ratio was achieved by using advanced spiral separators which will be discussed in this paper.

Keywords: chromite; beneficiation; wet high intensity magnetic separator; spiral concentrator; QEMSCAN; chromium-to-iron ratio

---

1. Introduction

Chromite is the only available source of chromium metal which is used for alloy steels in the form of ferrochrome. Production of ferroalloy is a high energy-intensive process and the economics of the process depends on the associated impurity level. Among the impurities, iron is one of the gangue element, which is present in the form of gangue minerals as well as in the chromite crystal lattice. It is well reported in the literature that the chromium-to-iron ratio (Cr:Fe ratio) of chromite ores plays a crucial role in the efficiency of ferrochrome production [1–3]. For removal of the gangue minerals, including iron-bearing minerals, beneficiation is mandatory prior to the smelting process [4–6]. Conventionally, the chromite ore is beneficiated by using gravity concentration techniques. With the decrease in the particle size, separation of the chromite particles by gravity separation becomes difficult due to the presence of near density gangue minerals as well as limitation in the equipment design [7]. Thus, the improvement in the Cr:Fe ratio of the product is limited with ferruginous chromite ore deposits compared to the siliceous type deposits. Beyond a specific limit, it is challenging to process these iron-bearing minerals by using conventional gravity separators [4,8,9].

Beneficiation flow sheets generally consist of different classifiers (mechanical screw classifiers, hydraulic based hindered settling classifiers as well as hydro cyclones), gravity units (jig, heavy media cyclones/separators, spiral concentrators as well as shaking tables) and dewatering processes (thickening and filtration). The ferruginous chromite ore deposits are mostly found in India. The achievable Cr:Fe ratio of the concentrates varies from 1.5 to 2.8 from a low-grade deposit with Cr:Fe ratios of 0.5 to 1.0 by using gravity separation [4,10–13].

There are few published data available about processing low-grade deposits or tailings in order to beneficiate chromite. Publications relevant to the gravity separation of different chromite deposits as well as tailing fraction were studied [10,14–20]. Most of these studies are limited to the flowsheet development, as well as recovery of the chromite particles. Gravity separators (water only cyclone, multi-gravity separator, and knelson concentrator) have also been used to improve the product quality of low-grade chromite ores as well as tailings of the beneficiation plant [12,16,17,21–24]. It should be added that the higher centrifugal Falcon separator has not been studied for chromite ores although this device is widely used in the mineral industry [25,26]. There are published papers available on magnetic separation as well as roasting assisted magnetic separation to discard the iron-bearing gangue minerals [9,13,27–31]. Flotation, selective flocculation, as well as magnetic carrier separation studies have also been carried out for different chromite ores to separate chromite from the gangue minerals [32–39]. However, most of the studied ores are either synthetic samples, or silicate rich gangue minerals. It should be mentioned that while separation of iron from chromite ore is reported in the literature, the separation feature from the concentrate produced by gravity separation is not available. Gravity concentration is a viable option due to the difference in the density of the associated gangue minerals and chromite. Also, some case studies discussed on the application of spiral concentrator for treating such ores from the gangue minerals. Furthermore, the spiral concentrator is widely used for separation of minerals due to its lower operating cost. The other option for treating such materials is by utilizing the difference between the magnetic susceptibility of the minerals. The wet high-intensity magnetic separator is a common piece of industrial equipment for treating ferrous minerals. In this case, the gangue minerals exhibit paramagnetic property due to their iron content. Therefore, any study in this direction will assist on the amenability for the enhancement of the Cr:Fe ratio from ferruginous chromite ores.

In this paper, the enhancement of the Cr:Fe ratio of a chromite ore was investigated by reprocessing via gravity separation and magnetic separation. A number of experiments were carried out by varying the critical variables of these units. The ore samples were thoroughly characterized by different techniques including automated advanced mineralogical tools.

2. Materials and Methods

2.1. Chromite Samples

Three different chromite samples were collected from a beneficiation plant as coarse, fine, and ultrafine concentrates. The chromite ore is from an open cast mine at the Sukinda region, India, and treated in a gravity-based plant. In the plant, the chromite ore was beneficiated by grinding to below 1 mm and classification to three different size fractions (by using hydrocyclone and hydraulic classifiers). Each fraction was subjected to three-stage separation (a combination of a rougher–scavenger–cleaner circuit) in spiral concentrators with different designs. The concentrate samples were filtered and dried and about 1 ton of each fractions was collected for this study.

2.2. Particle Characterisation

The particle size distribution (PSD) of the sample was measured in a laboratory sieve shaker. ICP-AES (Integra XL, I.R. Tech. Pvt. Ltd. (GBC Scientific Equipment, Victoria, Australia)) was carried out for the chemical assays. The mineral analysis was carried out by X-ray diffraction (XRD) supplied by PANanalytical B.V. (Malvern Panalytical, Almelo, The Netherlands) and QEMSCAN (FEI Company, Hillsboro, OR, United States). The details of these procedures are given in earlier publications [8,40].

2.3. Separation Processes

To enhance the Cr:Fe ratio of the samples, gravity concentration by the spiral concentrator and magnetic separation by the wet-high intensity magnetic separator was used. The details of the experimental setup along with the process conditions are discussed in Sections 2.3.1 and 2.3.2.

2.3.1. Spiral Concentrator Tests

Tests were carried out by varying the slurry flow rate and feed pulp density in two different spiral designs (HG10i and FM1 (Mineral Technologies, Carrara, Australia)). The other variables, such as splitter position, were maintained constant. Important parameters of the spiral concentrators and their design features along with the process conditions are given in Table 1. Experiments were carried out by varying the listed variables and operated in closed circuit arrangement whereby the product streams were recycled back to the feed tank. A valve located on the feed line was used to adjust the desired feed rate. The splitter position was maintained at 18 and 21 cm, respectively, for the HG10i and FM1 designs. The sampling was planned after setting the slurry flow rate with the designated slurry pulp density and it was allowed to run until a steady flow of the products streams achieved (i.e., more than 5 min). The products were collected, dried, weighed, and analyzed for the efficiency.

Table 1. The process parameters during the spiral concentrator tests.

Parameters	Spiral Design Types	
	HG10i	FM1
Spiral length (m)	2.84	3.12
Trough diameter (mm)	580	685
Trough height/pitch (mm)	350	360
Number of turns	8	8
Slurry flow rate (m^3/h)	0.9–2.3	0.9 and 2.3
Slurry pulp density (wt.%)	20 and 25	15, 20, 25, 30

2.3.2. Wet High Intensity Magnetic Concentrator Tests

Tests were carried out on a pilot-scale wet high-intensity magnetic separator (Jones P40 model WHIMS, MBE Coal & Minerals Technology GmbH, Gottfried-Hagen-Straße 20, Germany) by varying the magnetic field intensity at the different levels (0.4 to 1.35 T). The slurry pulp density and feed rate were maintained at 10% and 0.1 tph, respectively. Wash water flow rate of the WHIMS was also maintained at 7 L·min^{-1}. Also, the samples were also treated in two-stage at different magnetic field intensity. In the 1st stage, it was targeted to discard the non-magnetic particles at higher magnetic field intensity whereas paramagnetic minerals were targeted at a lower magnetic field intensity in the 2nd stage.

2.3.3. Experimental Analysis

After each experiment, the products (magnetic, middling, and non-magnetic for the WHIMS, and concentrate, middling, and tailing for the spiral concentrator) were collected for a fixed time, and dried. The weight distribution, as well as the elemental analysis for each product was measured and used for analyzing the grade, recovery, the Cr:Fe ratio and the Cr:Fe enrichment ratio. All tests were repeated at least 3 times, and the experimental error was found to be ±5%. The recovery and enrichment ratio for each test were calculated using Equations (1) and (2).

$$Recovery\ (\%)Cr_2O_3 = 100 \times \frac{Mass\ split \times Grade\ (\%)Cr_2O_3\ of\ the\ product}{Grade\ (\%)Cr_2O_3\ of\ the\ feed} \tag{1}$$

$$Cr:Fe\ enrichment\ ratio = \frac{Cr:Fe\ ratio\ of\ the\ product}{Cr:Fe\ ratio\ of\ the\ feed} \tag{2}$$

3. Result and Discussion

3.1. Characterisation Studies

3.1.1. Particle Size and Chemical Analysis

The PSD of the chromite concentrates is given in Figure 1. It is observed that 80% of the particles are below 715, 241, and 62 μm, respectively, for the coarse, fine, and ultrafine concentrates. Similarly, 50% of the particles are less than 345, 120, and 45μm, respectively, for the coarse, fine, and ultrafine concentrates. It is also observed that 21.2% of the sample is of ultrafine size (below 25 µm) for the ultrafine concentrate whereas this value is very minimum for the other two samples (i.e., 0.4% and 0.8% for the coarse and fine concentrates, respectively). The elemental analysis of the chromite concentrates is given in Table 2. Size by size chemical analysis results are also given in Table 3. The chromium oxide and total iron content of all size fractions in all concentrates are varied in a broader range. The iron content at finer fractions of all samples was found to be higher compared to the coarser size fractions.

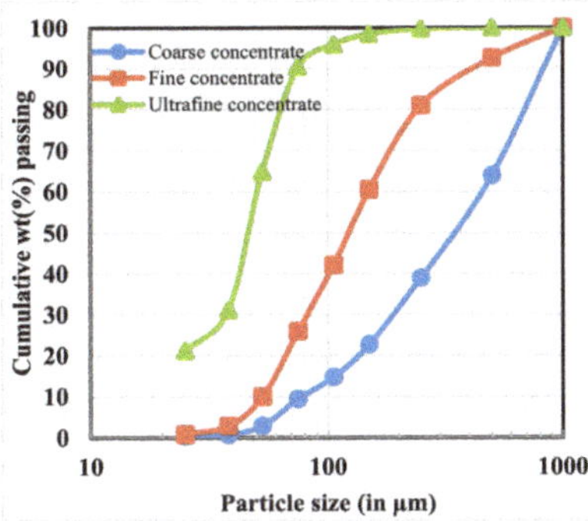

Figure 1. Particle size distribution of the chromite concentrates.

Table 2. Elemental assay of the concentrates.

Sample (Concentrates)	Assay (%)							
	$Fe_{(T)}$	CaO	SiO_2	MgO	Al_2O_3	Cr_2O_3	LOI	Cr:Fe Ratio
Coarse	14.4	0.3	7.9	9.3	9.5	49.9	3.8	2.4
Fine	13.2	0.3	3.9	10.8	9.2	53.7	3.7	2.8
Ultrafine	18.1	0.2	8.8	7.9	9.5	42.9	4.3	1.6

Table 3. Size by size elemental assay of the concentrates.

Size Fractions (µm)	Cr_2O_3	$Fe_{(T)}$	Cr:Fe Ratio
Coarse concentrate			
+710	51	14.6	2.4
−710 + 500	47.9	15.3	2.1
−500 + 250	48.3	15.1	2.2
−250 + 150	51.4	14.2	2.5
−150 + 75	51.8	12.1	2.9
−75	50.4	14.4	2.4
Fine concentrate			
+150	51.6	13.6	2.6
−150 + 75	54.6	13.1	2.9
−75 + 45	56.4	12.1	3.2
−45 + 25	55.8	13.4	2.8
−25	48.3	22.9	1.4
Ultrafine concentrate			
+150	25.6	19.8	0.9
−150 + 100	32.3	18.7	1.2
−100 + 75	33.1	18.6	1.2
−75 + 45	42.6	16.8	1.7
−45 + 25	51.8	14.4	2.5
−25	36.2	25.6	1.0

3.1.2. X-ray Diffraction Analysis

The XRD analysis results (Figure 2) shows that the samples contain mainly chromite magnesian ($Al_{0.06}Cr_{1.64}Fe_{0.78}Mg_{0.5}O_4Ti_{0.02}$), hematite ($Fe_2O_3$), goethite ($FeO(OH)$), spinel ($MgAl_2O_4$) and quartz($SiO_2$).

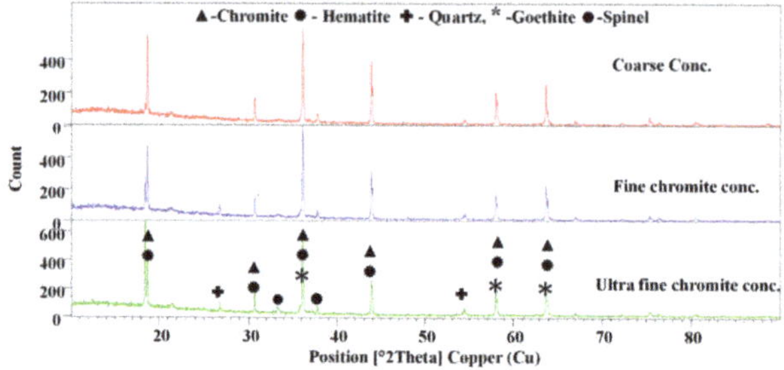

Figure 2. XRD pattern of the chromite concentrates.

3.1.3. Heavy Liquid Separation Tests

Heavy-liquid separation studies were carried out by using organic liquids to analyze the liberation of mineral grains of the samples. The heavy density separation study was performed by considering two types of density liquids of densities at 2.8 (bromoform) and 3.3 g/cm^3 (di-iodo-methane). Results are presented in Table 4. It is evident that the Cr:Fe ratio of the coarse concentrate increased from 2.4 to 2.7. Similarly, the Cr:Fe ratio of the fine and ultrafine concentrates were enhanced from 2.8 and 1.6, to 3.2 and 2.5, respectively. Results also indicate that the liberation of the iron-bearing minerals are significantly increasing when particle size decreases and therefore, the Cr:Fe ratio is enhanced to about 0.9 for the ultrafine size concentrate. Also, this is an indication on the amenability scope for the gravity separation for the removal of the iron-bearing gangue minerals. It can be assumed that by discarding the float parts to the non-magnetic fraction at higher magnetic intensity, the desired Cr:Fe ratio can be achieved.

Table 4. Results of the heavy liquid separation for chromite concentrates.

Product	Mass Split (%)	Assay (%)		Cr:Fe Ratio	Distribution (%)	
		Cr_2O_3	$Fe_{(T)}$		Cr_2O_3	$Fe_{(T)}$
Coarse concentrate						
Density (2.8 g/cm³)						
Float	7.5	11.6	11.3	0.7	1.7	5.9
Sink	92.5	53.0	14.7	2.5	98.3	94.1
Density (3.3 g/ cm³)						
Float	12.4	17.8	18.7	0.7	4.4	16.1
Sink	87.6	54.4	13.8	2.7	95.6	83.9
Fine concentrate						
Density (2.8 g/ cm³)						
Float	3.2	22.9	49.4	0.3	1.4	11.9
Sink	96.8	54.7	12.1	3.1	98.6	88.1
Density (3.3 g/ cm³)						
Float	6.1	10.7	33.3	0.2	1.2	15.3
Sink	93.9	56.5	12.0	3.2	98.8	84.7
Ultrafine concentrate						
Density (2.8 g/ cm³)						
Float	5.6	9.5	41.5	0.2	1.2	12.9
Sink	94.4	44.9	16.7	1.8	98.8	87.1
Density (3.3 g/ cm³)						
Float	14.3	10.5	48.9	0.1	3.5	38.7
Sink	85.7	48.3	13.0	2.5	96.5	61.3

3.1.4. Mineralogical Studies

The polished sections of the samples were studied by QEMSCAN. The results were derived by analyzing more than 8500 particle using five different polished blocks. The mineral composition of the concentrates are shown in Table 5. It can be seen that hematite, goethite, and iron silicates (serpentine, olivine group of minerals) are predominant as iron-bearing gangue minerals along with chromite.

Table 5. Mineral analysis of the concentrates along with the grain size of different minerals.

Minerals	Mineral Mass (%)			Grain Size (μm)		
	Coarse Concentrate	Fine Concentrate	Ultrafine Concentrate	Coarse Concentrate	Fine Concentrate	Ultrafine Concentrate
Chromite	81.0	89.4	76.1	110.7	58.5	42.1
Goethite	4.0	2.5	7.5	25.0	24.9	18.1
Hematite	4.1	2.3	4.0	26.7	35.9	20.8
Fe-silicate	4.7	1.8	5.1	19.9	21.0	14.4
Kaolinite	0.6	0.4	1.1	28.7	27.0	21.3
Silicate	1.0	0.8	1.6	28.6	36.7	24.3
Gibbsite	0.5	0.2	0.7	28.2	37.5	21.2
Others	4.1	2.7	4.0	17.5	25.7	15.7

Further, the average liberation grain size of the minerals are also given in Table 5. It is found that the average grain size for chromite and gangue minerals widely varies and the gangue minerals are liberated at a finer size compared to chromite. It is also observed that the gangue minerals are well liberated at ultrafine particle size ranges which indicates the desliming, as well as gravity separation techniques, can be used to separate.

While Figure 3 shows the presence of different minerals interlocked together but it is evident that a substantial amount of chromite is present in free form, as shown by the QEMSCAN images (Figure 4). In fact, 76.8%, 80.1% and 70.7% of the chromite particles are in free form in the coarse, fine, and ultrafine concentrates, respectively (which can be separated). Similarly, the association or interlock of the gangue minerals can be observed in Figure 3. For example, the Fe-silicate particles in the coarse concentrate indicate that 53.9% of the particles are interlocked with the chromite particles. In other words, the separation of Fe-silicate is possible at a higher efficiency due to their different density, but the interlocking of these particles with chromite enhances the apparent particle density and in turn, it minimizes the efficiency of the gravity separation. Similar interpretations are valid for different gangue minerals in these three cases. However, the enhancement of the Cr:Fe ratio due to the separation of liberated gangue particle may not help in decreasing the Fe content to zero. It should be noted that iron is present in the chromite spinel and it cannot be separated. Therefore, the Cr:Fe ratio data may not be helpful in this case. For a better understanding, the deportment of iron was analyzed by QEMSCAN. The elemental deportments for iron (Figure 5) showed that it is reported from hematite, goethite, and Fe-silicates along with chromite. In other words, a higher value of iron might have come from chromite particles as well as from the interlocking of iron-bearing minerals with chromite. This observation has been also reported earlier [9,28,41]. Therefore, the iron level in the concentrate can be lowered to only a certain extent, and beyond that, its separation is not feasible through the beneficiation. From Figure 5, it is observed that 72.4%, 87%, and 73% of the total iron is reported from chromite in the coarse, fine, and ultrafine concentrate, respectively. It is also found that iron is distributed mostly in hematite for the coarse concentrate, whereas Fe-silicate bearing minerals are dominant in the fine and ultrafine concentrate, respectively. It is concluded that the iron content in the beneficiation product can be lowered to 10.4%, 11.6%, and 13.2% for the coarse, fine, and ultrafine concentrate, respectively. Beyond these points, the iron content cannot be decreased further without losing chromite particles.

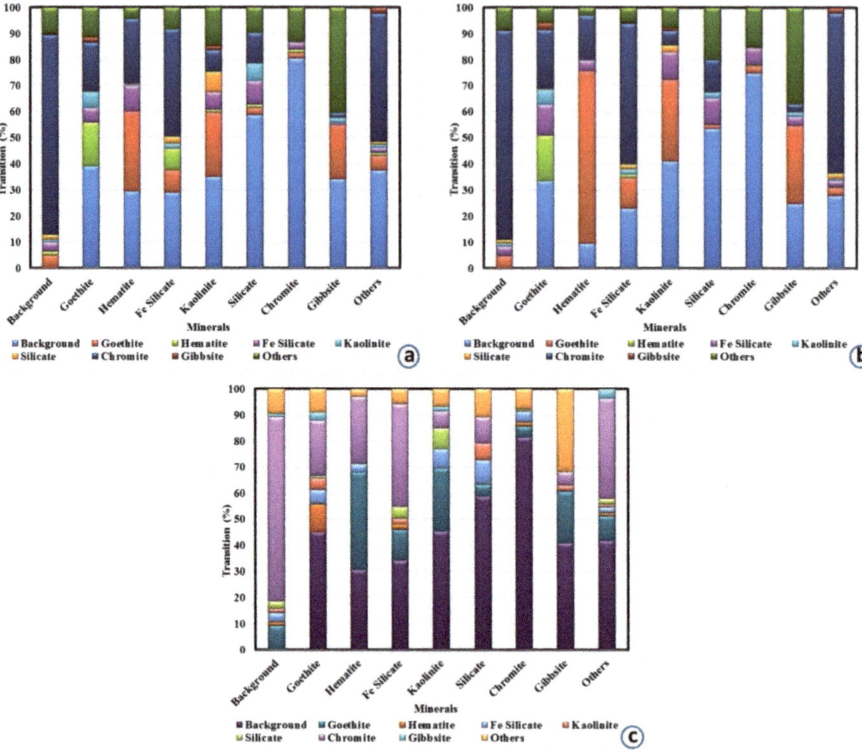

Figure 3. Association of different minerals in the chromite concentrate samples: (**a**) coarse, (**b**) fine, and (**c**) ultrafine concentrates.

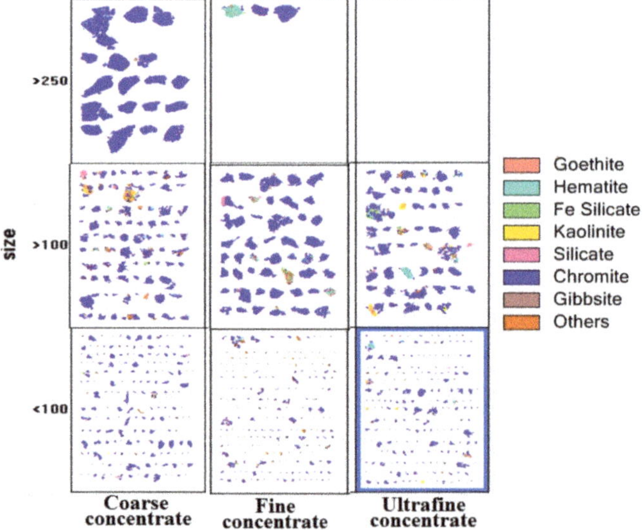

Figure 4. The QEMSCAN images of the samples in different size classes.

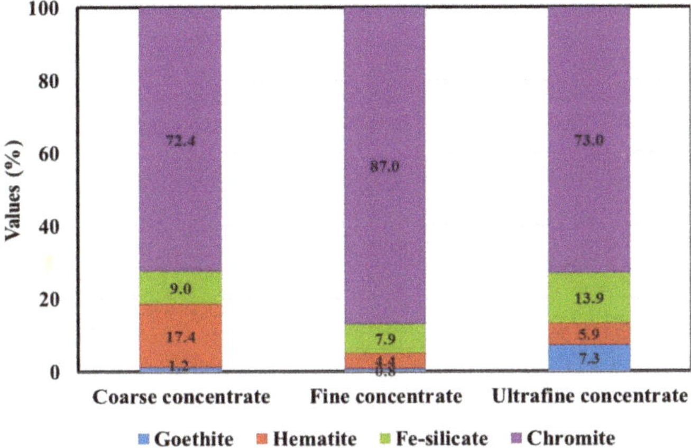

Figure 5. The iron deportment in different iron bearing mineral phases of the samples.

From the automated mineralogy, it is established that the best beneficiation strategy should decrease the hematite content of the coarse concentrate. The amenability of the separation is feasible by the gravity separation as well as magnetic separation. However, the magnetic separation may be more helpful due to the differences in the particle magnetic susceptibility. On the other hand, the gravity separation may be difficult due to the low concentration criterion between the minerals.

3.2. Separation Studies

3.2.1. Gravity Separation by Spiral Concentrator

Results of the separation using the FM1 spiral design for the coarse concentrate is shown in Figure 6. From Figure 6a, it is found that the grade of the concentrate decreases with an increase in the slurry pulp density at both slurry flow rates. It is also observed that the separation efficiency is poor with slurry pulp density higher than 20%. The recovery in the concentrate fraction is found to be in reverse trend of the grade, which is a general feature in the separation. Influence of the slurry flow rate also significantly influences the separation, which is evident from both grade and recovery values. A better separation is observed for the higher slurry flow rate of 2.3 m^3/h. A similar observation has been reported for other spiral designs for treating low-grade chromite and hematite ores [10,11,42]. The separation efficiency improves with increasing the feed velocity due to the enhancement of the centrifugal force acting on the particles. This, results in migration of coarser gangue particles (iron silicate bearing minerals) to the peripheral tailing stream. The influence of the slurry density, as well as flow rate on the separation, is analyzed in terms of Cr:Fe and enrichment ratio to understand the rejection of iron-bearing gangue minerals (Figure 6b). However, the rejection of iron-bearing gangue minerals is drastically affected with an increase in the slurry pulp density, but the separation is favorable at a higher slurry flow rate.

Similar tests were carried out in the HG10i spiral concentrator for the coarse concentrate sample. From Figure 7a, it is found that the optimum separation (grade) is at a pulp density of more than 20%. The layer of particles governs the separation inside the trough flows over the surface, which is dictated by the slurry flow rate as well as slurry density. The influence of slurry density on the coarse particle recovery is well explained in the literature [43,44]. However, the influence of the wash water flow rate is more of an influence than slurry density, which is not considered in the present research. It is also found that the grade, as well as the iron rejection, is better in the HG10i design. The lower diameter trough with a minimum pitch in the design facilitated the enhanced separation and resulted in Cr:Fe ratio values of more than 4 (Figure 7b).

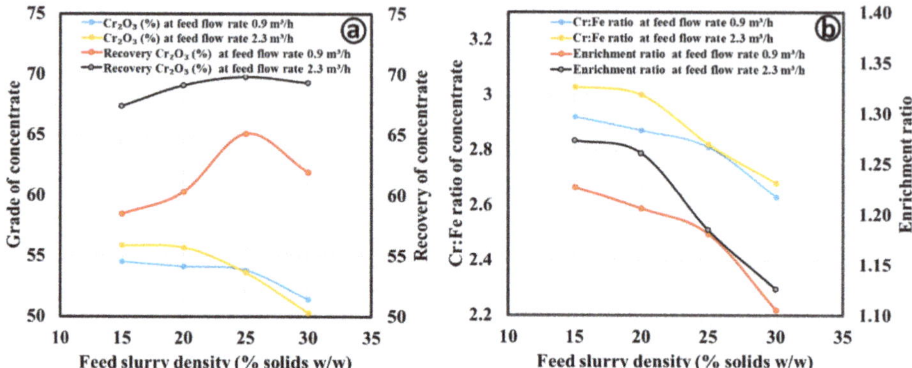

Figure 6. Influence of the feed flow rate and slurry density on the treatment of coarse concentrate by the spiral FM1 design (**a**) for grade and recovery; (**b**) for Cr:Fe ratio and enrichment ratio of concentrate.

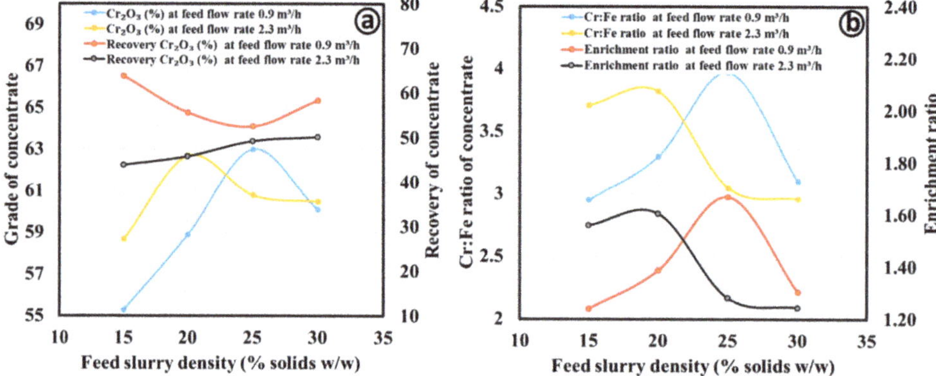

Figure 7. Influence of the feed flow rate and slurry density on the treatment of coarse concentrate by the spiral HG10i design (**a**) for grade and recovery; (**b**) for Cr:Fe ratio and enrichment ratio of concentrate.

Separation studies using fine concentrate samples were carried out in the spirals by varying slurry flow rate and pulp density. The results of the separation using FM1 spiral design is shown in Figure 8. From the Figure 8a, it is found that the grade of the concentrate increases with an increase in slurry pulp density up to 20% solids and after that, it decreases at the low slurry flow rate. The grade decreases with an increase of slurry density at the higher slurry flow rate. The Cr_2O_3 grade of the concentrate was enriched to values above 60% at the lower slurry flow rate. This is basically due to the better liberation of the chromite particles in the feed slurry along with a single layer of particle flow which enhances the separation. This in turn, results in better segregation of the heavy particles in the concentrate. Further, influence of the slurry density, as well as the flow rate on the separation was analyzed in terms of the Cr:Fe and enrichment ratios to understand the rejection of the iron-bearing gangue minerals. Results in Figure 8b shows that rejection of iron-bearing gangue minerals drastically decreases with an increase in slurry pulp density at the higher slurry flow rate. However, the separation is favorable at a lower slurry flow rate.

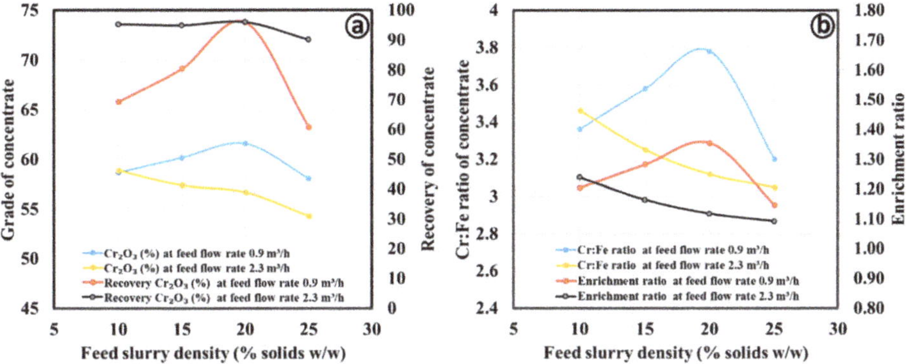

Figure 8. Influence of the feed flow rate and slurry density on the treatment of fine concentrate by the spiral FM1 design (**a**) for grade and recovery; (**b**) for Cr:Fe ratio and enrichment ratio of concentrate.

Figure 9 presents the results of similar tests carried out in the HG10i spiral design. Figure 9a indicates that the Cr_2O_3 grade of the concentrate increases with an increases in slurry pulp density at both flow rates. The optimum separation (grade) is also obtained at a pulp density of 25%. Generally, the separation inside the trough is governed by the layer of particles flows over the surface along with the hindered settling phenomena which enhances the separation based on the settling ratio of the heavy and light density particles. Similar to the coarse concentrate, the grade as well as iron rejection, were found to be better in the HG10i type. However, the recovery for both flow rates is below 70% while achieving a Cr_2O_3 grade higher than 60%. The Cr:Fe and enrichment ratios are found to be increasing as the slurry pulp density increases. The maximum Cr:Fe and enrichment ratios are observed for the higher slurry flow rate of 2.3 m^3/h (Figure 9b).

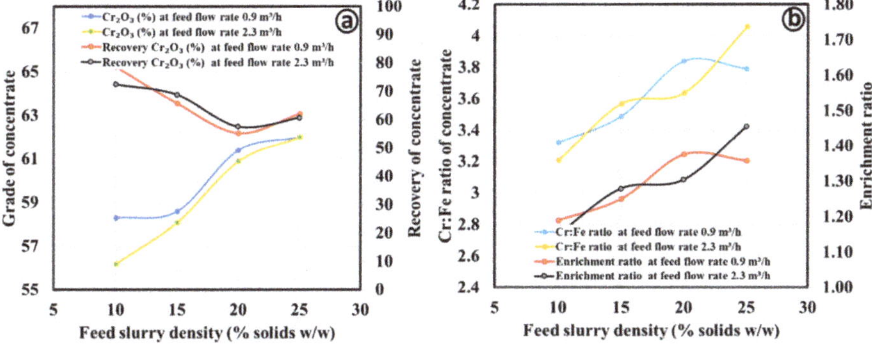

Figure 9. Influence of the feed flow rate and slurry density on the performance of spiral design HG10i in treating fine concentrate (**a**) for grade and recovery; (**b**) for Cr:Fe ratio and enrichment ratio of concentrate.

Results of the Cr_2O_3 upgradation using the FM1 spiral design are shown in Figure 10. From Figure 10a, it is found that the Cr_2O_3 grade of the concentrate increases with an increase in slurry pulp density up to 15% in both flow rates. The Cr_2O_3 grade of the concentrate was enriched but it does not exceed 55%. At the higher slurry flow rate, the grade does not exceed 50%. This is basically due to the entrainment of ultrafine gangue minerals to the concentrate flow. The change in the recovery during the separation is found to be the reverse trend of the grade, and the recovery is higher at lower slurry flow rates. Further, the influence of slurry density, as well as flow rate on the separation were analyzed in terms of Cr:Fe and enrichment ratios to understand the rejection of the iron-bearing gangue minerals. The rejection of iron-bearing gangue minerals is found to be proportional to the grade, but the separation is better at a lower slurry flow rate and pulp density of 15% (Figure 10b).

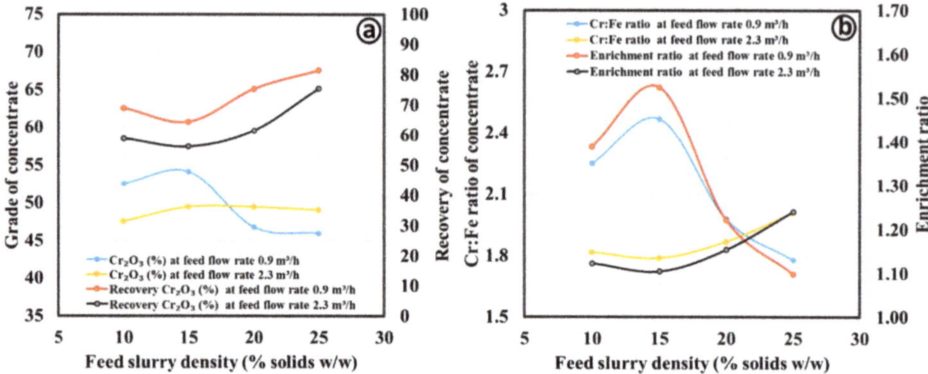

Figure 10. Influence of the feed flow rate and slurry density on the treatment of ultrafine concentrate by the spiral FM1design (**a**) for grade and recovery; (**b**) for Cr:Fe ratio and enrichment ratio of concentrate.

Figure 11 presents the results of similar tests carried out on the HG10i spiral design. Figure 11a indicates that the Cr_2O_3 grade of the concentrate increases with an increase in slurry pulp density up to 20%, and it decreases after that. However, there is insignificant change by varying the pulp density at higher flow rates. The recovery of Cr_2O_3 in the concentrate is found to be less than 40% in all tests. Therefore, the main disadvantage of this design is inadequate segregation of the ultrafine chromite particles. This is basically due to the inferior displacement of the particles radially due to the higher viscosity of the slurry. The Cr:Fe and enrichment ratios (Figure 11b) are also increasing when the slurry pulp density increases to 20%, and after that it decreases. The maximum Cr:Fe and enrichment ratio values are observed for the higher slurry flow rate of 2.3 m^3/h.

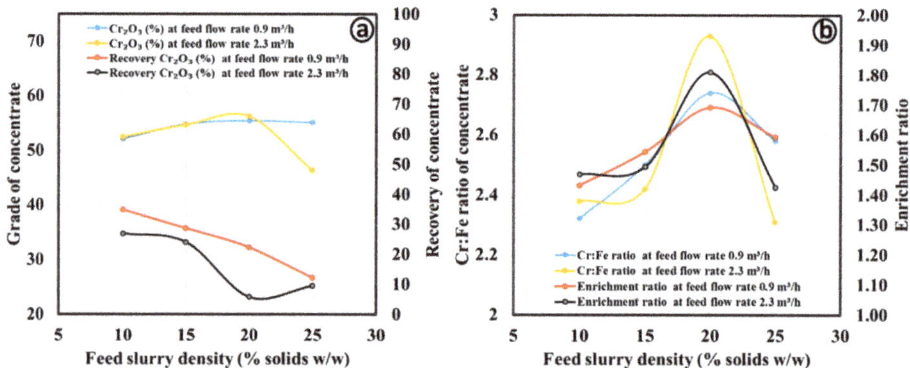

Figure 11. Influence of the feed flow rate and slurry density on the treatment of ultrafine concentrate by the spiral HG10i design (**a**) for grade and recovery; (**b**) for Cr:Fe ratio and enrichment ratio of concentrate.

3.2.2. Magnetic Separation by Wet High Intensity Magnetic Separator

Magnetic separation in WHIMS was carried out by varying the magnetic field intensity from 0.4 to 1.3 T. Figure 12a shows that there is an increase in the grade of the magnetic fraction of the coarse concentrate with an increase in magnetic field intensity up to 1.1T, and it decreases drastically after that. This is due to the paramagnetic nature of the chromite and rejection of the silicate bearing gangue minerals in the non-magnetic fraction. However, at a magnetic field intensity of 1.3 T, there may be the attraction of goethite and iron-silicate bearing minerals which are paramagnetic and are reported to the magnetic fraction. Similarly, the iron rejection and the efficiency of separation with the change in the magnetic field intensity is shown in Figure 12b. The separation of iron-bearing minerals from chromite is found to be limited as the Cr:Fe ratio of the product is below 2.7. This is due to the abundance of the near magnetic susceptibility minerals as well their poor liberation.

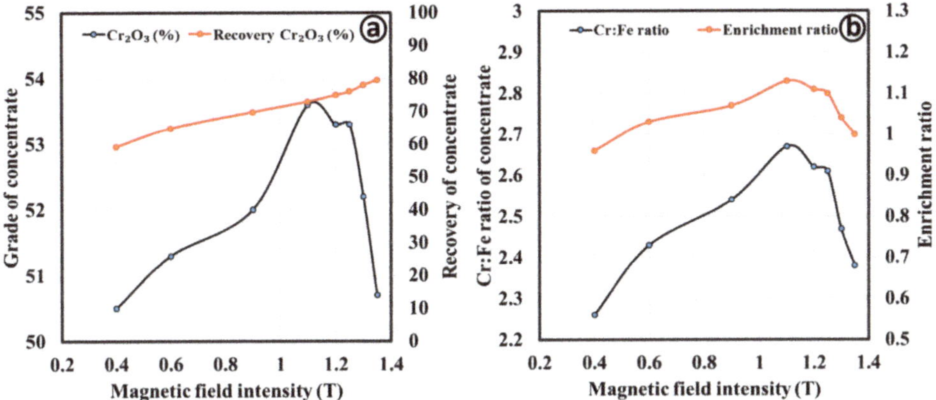

Figure 12. Influence of the magnetic field intensity on the treating coarse concentrate by the WHIMS (**a**) for grade and recovery; (**b**) for Cr:Fe ratio and enrichment ratio of concentrate.

Similarly, the results for the fine and ultrafine concentrates are given in Figures 13 and 14. Figure 13 shows separation of chromite, as well as the rejection of iron-bearing minerals at a magnetic field intensity of 1.2 T. Further, the Cr_2O_3 grade of the magnetic fraction is enriched to 59.7% from 53.7% in the feed. The Cr-Fe ratio was also increased from 2.8 to 3.94.

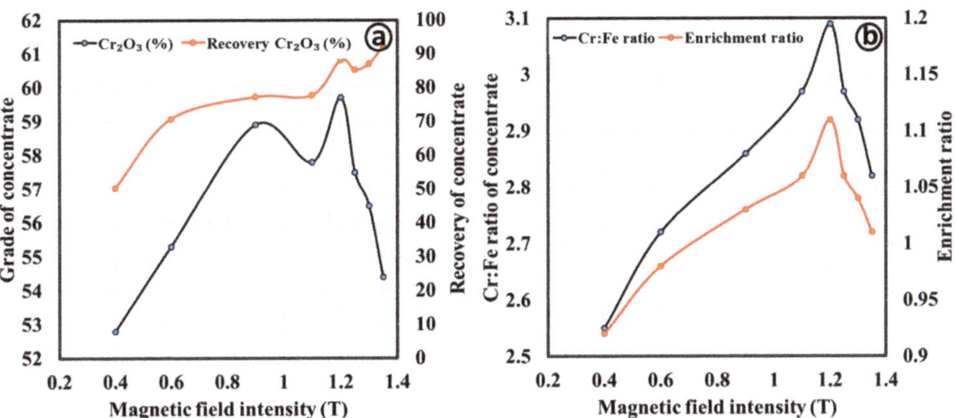

Figure 13. Influence of the magnetic field intensity on the treating fine concentrate by the WHIMS (**a**) for grade and recovery; (**b**) for Cr:Fe ratio and enrichment ratio of concentrate.

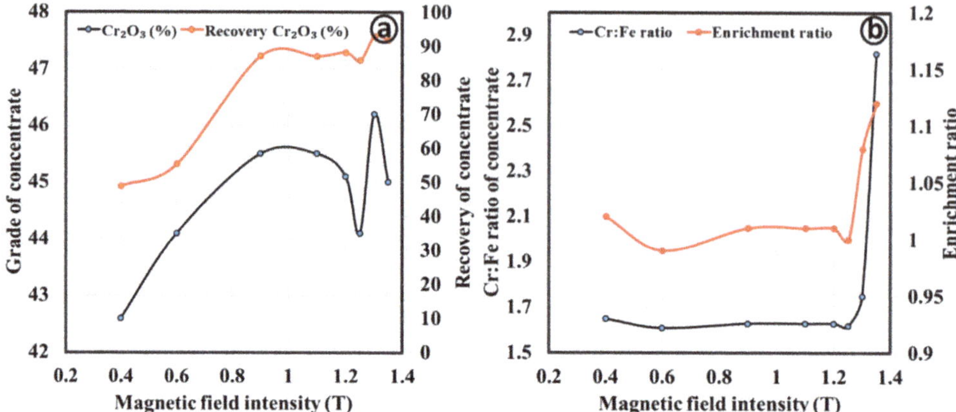

Figure 14. Influence of the magnetic field intensity on the treating ultrafine concentrate by the WHIMS (**a**) for grade and recovery; (**b**) for Cr:Fe ratio and enrichment ratio of concentrate).

The maximum achievable Cr_2O_3 grade for the ultrafine concentrate is 45% from a feed with 42.9% Cr_2O_3 (Figure 14), with enhancement of the Cr:Fe ratio from 1.6 to 1.8. The maximum grade is reported at a magnetic field intensity of 0.9 T. The separation of the iron-bearing particles is found to be minimal for the ultrafine concentrate compared to other two samples. This is mainly due to inefficient capture of the magnetic particles at finer particle size range as well as the abundance of hematite and goethite in the sample which are paramagnetic and are reported along with chromite.

3.3. Discussion

From Figure 15, it is observed that the WHIMS can be used for increasing chromium oxide recovery, but it does not result in a significant Cr_2O_3 grade and Cr:Fe ratio. Further, the two-stage separation was adopted to discard hematite and goethite at a lower magnetic field intensity of 0.4 T and iron-bearing silicates at 1.2 T. The two-stage separation for such low-grade ferruginous chromite ore is well reported in the literature [13]. The results of the two-stage separation are found to be insignificant as there is an incremental change in the grade, but the recovery decreases drastically. However, the spiral concentrator can be used to achieve products with a Cr:Fe ratio higher than 2.8, and recovery level at 70%. In the case of the fine concentrate sample, the grade–recovery relationship (Figure 16a) is found to be identical in both units. However, the enhancement of the Cr:Fe ratio is weak in the WHIMS, compared to the spiral separator (Figure 16b,c). This is due to the efficient separation of the non-magnetic gangue minerals (e.g., iron-bearing silicates, quartz) from chromite whereas an inefficient separation of hematite and goethite. Chromite has been previously separated at magnetic field intensities above 1 T by discarding the gangue minerals (serpentine and olivine, which associate with chromite) [32]. This is possible since chromite has slightly more magnetic susceptibility than gangue minerals due to its higher iron content. The major gangue minerals in our case are hematite and goethite along with the silicate bearing phases which hindered the efficient separation as their magnetic susceptibility are similar. The trend of correlation for the separation for ultrafine concentrate (Figure 17) is found to be identical with the coarse concentrate. In all samples, the spiral concentrate is found to be more efficient for the enhancement of the Cr:Fe ratio compared to the WHIMS. Also, it is evident that the separation is primarily influenced by the spiral design.

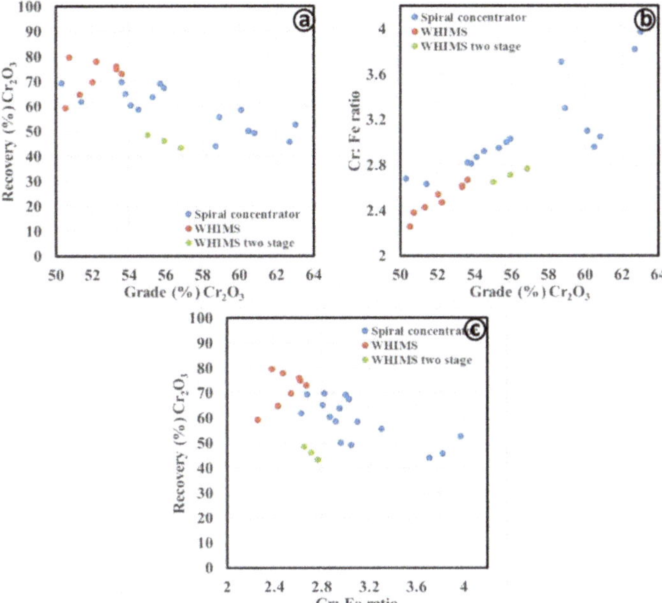

Figure 15. Comparison of the performance of the spiral concentrator and the WHIMS in treating coarse concentrate (**a**) for grade and recovery; (**b**) for grade and Cr:Fe rato; (**c**) for Cr:Fe rato and recovery.

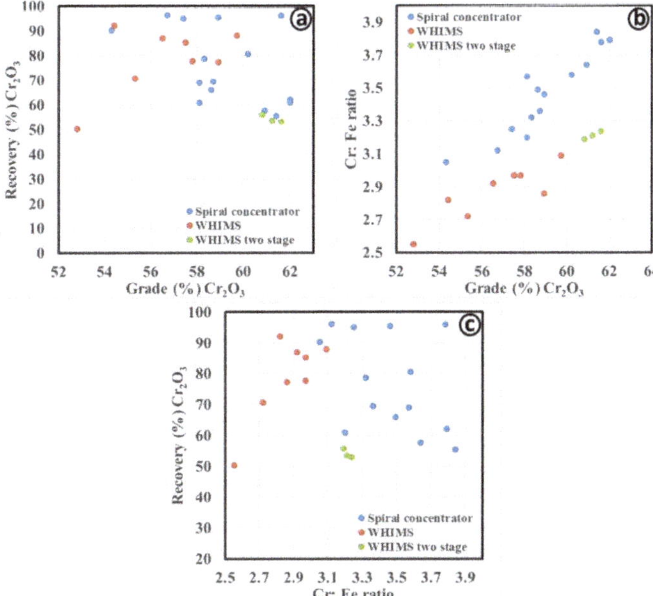

Figure 16. Comparison of the performance of the spiral concentrator and the WHIMS in treating fine concentrate (**a**) for grade and recovery; (**b**) for grade and Cr:Fe rato; (**c**) for Cr:Fe rato and recovery.

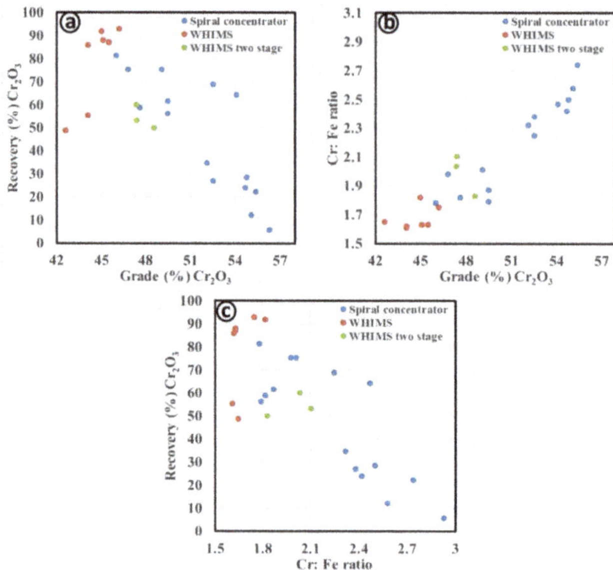

Figure 17. Comparison of the performance of the spiral concentrator and the WHIMS in treating ultrafine concentrate (**a**) for grade and recovery; (**b**) for grade and Cr:Fe rato; (**c**) for Cr:Fe rato and recovery.

The optimum conditions derived from the spiral concentrator is given in Table 6. The Cr:Fe ratio of the fine concentrate is upgraded from 2.38 to 3.97 with a Cr_2O_3 grade of 63% and recovery of 52.5% in the HG10i spiral. Upgradation of the Cr:Fe ratio with higher recovery is possible with two-stage separation. Similarly, the Cr:Fe ratio of the fine concentrate was upgraded from 2.79 to 4.06, with a Cr_2O_3 grade of 62.0% and 52.6% recovery. Optimum condition was obtained at a feed rate of 2.3 m^3/h and pulp density of 25% in HG10i spiral. Also, a higher Cr:Fe ratio was achieved in the FM1 design (3.78 vs. 2.79) with a Cr_2O_3 grade of 61.6% and mass recovery of 83.7% at a feed rate of 0.9 m^3/h and pulp density of 20%. The lower feed flow rate is favorable while separating the ultrafine particles of less than 45 μm [10]. Further upgradation the Cr:Fe ratio is possible with two stages of separation with a combination of these two spiral designs. The Cr:Fe ratio of the ultrafine concentrate was upgraded to 2.47 from 1.62 with a Cr_2O_3 grade of 54.1% and 51.1% recovery. This was achieved at a feed rate of 0.9 m^3/h and pulp density of 15% in the FM1 spiral.

Table 6. Results of the optimum condition for enhancing Cr:Fe ratios.

Process Parameters	Products	Mass Split (%)	Cr_2O_3 (%)	Cr:Fe Ratio	Cr_2O_3 (%) Recovery	Enrichment Ratio
		Coarse concentrate				
Spiral type: HG10i; Slurry density	Concentrate	41.6	63.0	3.97	52.5	1.67
(wt.%): 25; Slurry flow rate (m^3/h); 0.9;	Middling	15.9	49.4	2.34	15.7	0.98
Splitter position: 20 cm	Tailing	42.6	37.3	1.43	31.8	0.60
		Fine concentrate				
Spiral type: HG10i; Slurry density	Concentrate	52.6	62.0	4.06	60.8	1.46
(wt.%): 25; Slurry flow rate (m^3/h); 2.3;	Middling	29.2	57.6	3.17	31.3	1.14
Splitter position: 20 cm	Tailing	18.2	23.2	0.72	7.8	0.26
		Ultrafine concentrate				
Spiral type: FM1; Slurry density (wt.%):	Concentrate	51.1	54.1	2.47	64.4	1.52
15; Slurry flow rate (m^3/h); 0.9; Splitter	Middling	24.1	27.5	0.94	15.5	0.58
position: 18 cm	Tailing	24.8	34.9	1.05	20.2	0.65

The optimum conditions for the two-stage separation in WHIMS are given in Tables 7–9. The Cr:Fe ratio of the coarse concentrate was upgraded to 2.62 from 2.38 with a Cr_2O_3 grade of 53.3% and mass recovery of 70.3%. Further upgradation the Cr:Fe ratio is possible to 2.77 with a mass recovery of 38.1% and Cr_2O_3 grade of 56.8% with two-stage separation. Similarly, Table 8 shows that the Cr:Fe ratio of the fine concentrate was upgraded to 3.09 from 2.79 with a Cr_2O_3 grade of 59.7% and mass recovery of 79.3%. Further upgradation of Cr:Fe ratio is possible to 3.24 with a mass recovery of 46.1% and Cr_2O_3 grade of 61.6% with two-stage separation. The Cr:Fe ratio of the ultrafine concentrate was upgraded to 1.81 from 1.62 with a Cr_2O_3 grade of 45% and yield of 87.4% (Table 9). Further upgradation of the Cr:Fe ratio is possible to 2.11 with a yield of 48.1% and Cr_2O_3 grade of 47.4% with the two-stage separation.

Table 7. Results of the two-stage separation in the WHIMS for the coarse concentrate.

Magnetic Field Intensity	Products	Mass Split (%)	Cr_2O_3 (%)	Cr:Fe Ratio	Cr_2O_3 (%) Recovery	Enrichment Ratio
	Magnetic	70.3	53.3	2.6	75.0	1.11
1.2	Middling	8.3	46.4	2.1	7.7	0.89
	Nonmagnetic	21.4	40.2	1.8	17.3	0.74
	Magnetic	38.1	56.8	2.8	43.4	1.17
0.4	Middling	12.5	52.74	2.7	13.2	1.12
	Nonmagnetic	19.7	46.76	2.3	18.5	0.98

Table 8. Results of the two-stage separation in the WHIMS for the fine concentrate.

Magnetic Field Intensity	Products	Mass Split (%)	Cr_2O_3 (%)	Cr:Fe Ratio	Cr_2O_3 (%) Recovery	Enrichment Ratio
	Magnetic	79.3	59.7	3.1	88.1	1.11
1.2	Middling	4.0	48.3	2.6	3.6	0.92
	Nonmagnetic	16.6	26.4	1.4	8.2	0.50
	Magnetic	46.1	61.6	3.2	52.9	1.16
0.4	Middling	11.4	57.6	3.2	12.2	1.14
	Nonmagnetic	21.8	56.6	2.7	23.0	0.98

Table 9. Results of the two-stage separation in the WHIMS for the ultrafine concentrate.

Magnetic Field Intensity	Products	Mass Split (%)	Cr_2O_3 (%)	Cr:Fe Ratio	Cr_2O_3 (%) Recovery	Enrichment Ratio
	Magnetic	87.4	45.0	1.8	91.8	1.12
1.3	Middling	4.1	37.3	2.0	3.6	1.21
	Nonmagnetic	8.5	20.2	1.1	4.0	0.70
	Magnetic	48.1	47.4	2.1	53.2	1.30
0.4	Middling	15.0	45.3	1.6	15.8	0.99
	Nonmagnetic	24.3	40.3	1.5	22.8	0.91

4. Conclusions

The chromite fine particles generated from a gravity-based beneficiation plant was reprocessed using gravity (spiral concentrator) and magnetic separation (WHIMS) to enhance the Cr:Fe ratio. Initially, a detailed mineralogical characterization was carried out to evaluate the feasibility of the separation process. The Cr:Fe ratio of the fine concentrate sample was found to be higher compared to the other two samples (i.e., 2.8 compared to 2.4 and 1.6 for the coarse and ultrafine concentrates, respectively). It was also found that the Cr_2O_3 content is more segregated at the intermediate size fraction in all samples. XRD data revealed the abundance of iron-bearing minerals (goethite, hematite) and quartz along with chromite. QEMSCAN also showed that 76.8%, 80.1%, and 70.7% of the chromite particles are liberated in the coarse, fine, and ultrafine concentrates, respectively. About 72.4%, 87%, and 73% of the total iron was also found to be reported from chromite in the coarse, fine, and ultrafine concentrates, respectively.

Separation of the iron-bearing gangue minerals was found to be more efficient using gravity concentration rather than magnetic separation. It was also found that the spiral design plays an important role on the efficient chromite segregation. The HG10i model was found to be effective for the separation of chromite in processing coarse and fine concentrate samples, whereas FM1 was found to be suitable for the ultrafine concentrate sample. The Cr:Fe ratio of the coarse, fine, and ultrafine concentrate samples was enhanced to 3.97, 4.06, and 2.47, respectively, via gravity separation. Further upgradation of the Cr:Fe ratio is possible with two or three stages of separation using this type of spiral designs. Also, the recovery values may be further enhanced by recirculating the middling fractions of the spiral concentrator, as well as optimizing the multistage separation in a rougher–scavenger–cleaner circuit. In addition, future studies can focus on the influence of wash water with different trough design or number of troughs, along with other process variables affecting the separation performance. It is also worth suggesting a detailed investigation of enhancing the Cr:Fe ratio of the chromite ore by treatment in the recently developed enhanced gravity separators (e.g., water only cyclone, multi-gravity separator, knelson concentrator, and falcon concentrator).

Author Contributions: For research articles with several authors, a short paragraph specifying their individual contributions must be provided. The following statements should be used "conceptualization, S.K.T and Y.R.M; methodology, S.K.T.; software, V.S.; validation, S.K.T., Y.R.M. and V.S.; formal analysis, S.K.T.; investigation, S.K.T. and Y.R.M.; resources, S.K.T.; data curation, S.K.T.; writing—original draft preparation, S.K.T.; writing—review and editing, S.F., L.O.F.; visualization, S.F.; supervision, L.O.F.; project administration, S.K.T.; funding acquisition, S.K.T.", please turn to the CRediT taxonomy for the term explanation. Authorship must be limited to those who have contributed substantially to the work reported.

Funding: This research received no external funding.

Acknowledgments: The authors are thankful to the management of Tata Steel Ltd. for the support and permission to publish this study. SKT would like to acknowledge Labex Resources21 supported by the French National Research Agency through the national program "Investissements d'Avenir" [reference ANR–10–LABX–21] for his fellowship.

Conflicts of Interest: The authors declare no conflict of interest.

References

1. Azari, J. Effect of Chrome ore Qulaity on Ferrochrome Production Efficiency. In Proceedings of the Tenth International Ferroalloys Congress, Cape Town, South Africa, 1—4 February 2004.
2. Dwarapudi, S.; Tathavadkar, V.; Rao, B.C.; Kumar, T.K.S.; Ghosh, T.K.; Denys, M. Development of Cold Bonded Chromite Pellets for Ferrochrome Production in Submerged Arc Furnace. *ISIJ Int.* **2013**, *53*, 9–17. [CrossRef]
3. Nurjaman, F.; Subandrio, S.; Ferdian, D.; Suharno, B. Effect of basicity on beneficiated chromite sand smelting process using submerged arc furnace. *AIP Conf. Proc.* **2018**, *1964*, 020009.
4. Murthy, Y.R.; Tripathy, S.K.; Kumar, C.R. Chrome ore beneficiation challenges & opportunities–A review. *Miner. Eng.* **2011**, *24*, 375–380.
5. Nafziger, R.H. A review of the deposits and beneficiation of lower-grade chromite. *J. S. Afr. Inst. Min. Metall.* **1982**, *82*, 205–226.
6. Atalay, U.; Özbayoğlu, G. Beneficiation and agglomeration of chromite—It's application in Turkey. *Miner. Process. Extr. Metall. Rev.* **1992**, *9*, 185–194. [CrossRef]
7. Das, A.; Sarkar, B. Advanced gravity concentration of fine particles: A review. *Miner. Process. Extr. Metall. Rev.* **2018**, *39*, 359–394. [CrossRef]
8. Tripathy, S.K.; Murthy, Y.R.; Singh, V. Characterisation and separation studies of Indian chromite beneficiation plant tailing. *Int. J. Miner. Process.* **2013**, *122*, 47–53. [CrossRef]
9. Tripathy, S.K.; Banerjee, P.; Suresh, N. Magnetic separation studies on ferruginous chromite fine to enhance Cr: Fe ratio. *Int. J. Miner. Metall. Mater.* **2015**, *22*, 217–224. [CrossRef]
10. Tripathy, S.K.; Murthy, Y.R. Modeling and optimization of spiral concentrator for separation of ultrafine chromite. *Powder Technol.* **2012**, *221*, 387–394. [CrossRef]

11. Tripathy, S.K.; Murthy, Y.R. Multiobjective optimisation of spiral concentrator for separation of ultrafine chromite. *Int. J. Min. Miner. Eng.* **2012**, *4*, 151–162. [CrossRef]

12. Sunil, K.T.; Rama, M.Y.; Tathavadkar, V.; Mark, B.D. Efficacy of multi gravity separator for concentrating ferruginous chromite fines. *J. Min. Metall. Min.* **2012**, *48*, 39–49.

13. Tripathy, S.K.; Murthy, Y.R.; Singh, V.; Suresh, N. Processing of Ferruginous Chromite Ore by Dry High-Intensity Magnetic Separation. *Miner. Process. Extr. Metall. Rev.* **2016**, *37*, 196–210. [CrossRef]

14. Özgen, S. Modelling and optimization of clean chromite production from fine chromite tailings by a combination of multigravity separator and hydrocyclone. *J. S. Afr. Inst. Min. Metall.* **2012**, *112*, 387–394.

15. Öztürk, F.D.; Temel, H.A. Beneficiation of Konya-Beyşehir Chromite for Producing Concentrates Suitable for Industry. *JOM* **2016**, *68*, 2449–2454. [CrossRef]

16. Çiçek, T.; Cengizler, H.; Cöcen, İ. An efficient process for the beneficiation of a low grade chromite ore. *Miner. Process. Extr. Metall.* **2010**, *119*, 142–146. [CrossRef]

17. Özgen, S. Clean chromite production from fine chromite tailings by combination of Multi Gravity Separator and Hydrocyclone. *Sep. Sci. Technol.* **2012**, *47*, 1948–1956. [CrossRef]

18. Singh, R.K.; Dey, S.; Mohanta, M.K.; Das, A. Enhancing the Utilization Potential of a Low Grade Chromite Ore through Extensive Physical Separation. *Sep. Sci. Technol.* **2014**, *49*, 1937–1945. [CrossRef]

19. Can, İ.B.; Özsoy, B.; Ergün, Ş.L. Developing an optimum beneficiation route for a low-grade chromite ore. *Physicochem. Probl. Miner. Process.* **2019**, *55*, 865–878.

20. Tripathy, S.K.; Ramamurthy, Y.; Singh, V. Recovery of chromite values from plant tailings by gravity concentration. *J. Miner. Mater. Charact. Eng.* **2011**, *10*, 13. [CrossRef]

21. Akar Sen, G. Application of Full Factorial Experimental Design and Response Surface Methodology for Chromite Beneficiation by Knelson Concentrator. *Minerals* **2016**, *6*, 5. [CrossRef]

22. Çiçek, T.; Cöcen, I.; Engin, V.T.; Cengizler, H.; Şen, S. Technical and economical applicability study of centrifugal force gravity separator (MGS) to Kef chromite concentration plant. *Miner. Process. Extr. Metall.* **2008**, *117*, 248–255. [CrossRef]

23. Tripathy, S.K.; Bhoja, S.K.; Murthy, Y.R. Processing of chromite ultra-fines in a water only cyclone. *Int. J. Min. Sci. Technol.* **2017**, *27*, 1057–1063. [CrossRef]

24. Aslan, N. Application of response surface methodology and central composite rotatable design for modeling and optimization of a multi-gravity separator for chromite concentration. *Powder Technol.* **2008**, *185*, 80–86. [CrossRef]

25. Foucaud, Y.; Dehaine, Q.; Filippov, L.O.; Filippova, I.V. Application of Falcon Centrifuge as a Cleaner Alternative for Complex Tungsten Ore Processing. *Minerals* **2019**, *9*, 448. [CrossRef]

26. Farrokhpay, S.; Filippov, L.; Fornasiero, D. Pre-concentration of nickel in laterite ores using physical separation methods. *Miner. Eng.* **2019**, *141*, 105892. [CrossRef]

27. Altın, G.; İnal, S.; Alp, İ. Recovery of Chromite from Processing Plant Tailing by Vertical Ring and Pulsating High-Gradient Magnetic Separation. *MT Bilimsel* **2018**, *13*, 23–35.

28. Ayinla, K.I.; Baba, A.A.; Tripathy, B.C.; Ghosh, M.K.; Dwari, R.K.; Padhy, S.K. Enrichment of a Nigerian chromite ore for metallurgical application by dense medium flotation and magnetic separation. *Metall. Res. Technol.* **2019**, *116*, 324. [CrossRef]

29. Subandrio, S.; Dahani, W.; Alghifar, M.; Purwiyono, T.T. Enrichment Chromite Sand Grade Using Magnetic Separator. *IOP Conf. Ser. Mater. Sci. Eng.* **2019**, *588*, 012033. [CrossRef]

30. Gupta, P.; Bhandary, A.K.; Chaudhuri, M.G.; Mukherjee, S.; Dey, R. Kinetic Studies on the Reduction of Iron Oxides in Low-Grade Chromite Ore by Coke Fines for Its Beneficiation. *Arab. J. Sci. Eng.* **2018**, *43*, 6143–6154. [CrossRef]

31. Tripathy, S.K.; Singh, V.; Ramamurthy, Y. Improvement in Cr:Fe Ratio of Indian Chromite Ore for Ferro Chrome Production. *Int. J. Min. Eng. Miner. Process.* **2012**, *1*, 101–106.

32. Güney, A.; Onal, G.; Çelik, M.S. A new flowsheet for processing chromite fines by column flotation and the collector adsorption mechanism. *Miner. Eng.* **1999**, *12*, 1041–1049. [CrossRef]

33. Ucbas, Y.; Bozkurt, V.; Bilir, K.; Ipek, H. Concentration of chromite by means of magnetic carrier using sodium oleate and other reagents. *Physicochem. Probl. Miner. Process.* **2014**, *50*, 767–782.

34. Ucbas, Y.; Bozkurt, V.; Bilir, K.; Ipek, H. Separation of Chromite from Serpentine in Fine Sizes using Magnetic Carrier. *Sep. Sci. Technol.* **2014**, *49*, 946–956. [CrossRef]

35. Gallios, G.P.; Deliyanni, E.A.; Peleka, E.N.; Matis, K.A. Flotation of chromite and serpentine. *Sep. Purif. Technol.* **2007**, *55*, 232–237. [CrossRef]

36. Seifelnasr, A.A.; Tammam, T. Flotation behavior of Sudanese chromite ores. *J. Eng. Sci. Fac. Eng. Assiut Univ.* **2011**, *39*, 649–661.

37. Alesse, V.; Belardi, G.; Freund, J.; Piga, L.; Shehu, N. Acidic medium flotation separation of chromite from olivine and serpentine. *Min. Metall. Explor.* **1997**, *14*, 26–35. [CrossRef]

38. Güney, A.; Atak, S. Separation of chromite from olivine by anionic collectors. *Fizykochem. Probl. Miner.* **1997**, *31*, 99–106.

39. Panda, L.; Banerjee, P.K.; Biswal, S.K.; Venugopal, R.; Mandre, N.R. Modelling and optimization of process parameters for beneficiation of ultrafine chromite particles by selective flocculation. *Sep. Purif. Technol.* **2014**, *132*, 666–673. [CrossRef]

40. Devasahayam, S. Predicting the liberation of sulfide minerals using the breakage distribution function. *Miner. Process. Extr. Metall.* **2015**, *36*, 136–144. [CrossRef]

41. Das, S.K. Quantitative mineralogical characterization of chrome ore beneficiation plant tailing and its beneficiated products. *Int. J. Miner. Metall. Mater.* **2015**, *22*, 335–345. [CrossRef]

42. Dixit, P.; Tiwari, R.; Mukherjee, A.K.; Banerjee, P.K. Application of response surface methodology for modeling and optimization of spiral separator for processing of iron ore slime. *Powder Technol.* **2015**, *275*, 105–112. [CrossRef]

43. Dehaine, Q.; Filippov, L.O. Modelling heavy and gangue mineral size recovery curves using the spiral concentration of heavy minerals from kaolin residues. *Powder Technol.* **2016**, *292*, 331–341. [CrossRef]

44. Sadeghi, M.; Bazin, C.; Renaud, M. Effect of wash water on the mineral size recovery curves in a spiral concentrator used for iron ore processing. *Int. J. Miner. Process.* **2014**, *129*, 22–26. [CrossRef]

Article

Selective Capture of Magnetic Wires to Particles in High Gradient Magnetic Separation

Jianwu Zeng, Xiong Tong *, Fan Yi and Luzheng Chen *

Faculty of Land Resource Engineering, Kunming University of Science and Technology, Kunming 650093, China
* Correspondence: xiongtong2000@yahoo.com (X.T.); chluzheng@kmust.edu.cn (L.C.);
   Tel.: +86-871-6518-0062 (L.C.)

Received: 30 June 2019; Accepted: 21 August 2019; Published: 23 August 2019

Abstract: High gradient magnetic separation (HGMS) achieves effective separation to fine weakly magnetic minerals using numerous small magnetic wires in matrix, and its separation performance is inherently dependent on the capture characteristics of the wires. In this work, the selective capture of magnetic wire to particles in high gradient magnetic field was theoretically described and simulated using COMSOL Multiphysics. It was found that the capture trajectories of a small amount of particles under the ideal condition was significantly different from those of a large amount of particles under the actual condition, and non-magnetic particles would be much more easily entrained into magnetic deposits captured onto the wire surface under the actual condition than those under the ideal condition. These theoretical and simulated results were basically validated with the experimental magnetic capture to an ilmenite ore, and the wires in slow feed mode have achieved much higher capture selectivity than those in the fast feed mode. For instance, at the magnetic induction of 0.8 T, the TiO$_2$ grade of magnetic deposits captured by 3 mm diameter wire in the slow feed model reached 36.78%, which is higher than 28.32% in the fast feed model. The selective capture difference between the fast and slow feed models increased with increase in the magnetic induction and with decrease in the pulsating frequency. This investigation contributes to improve HGMS performance in concentrating fine weakly magnetic ores.

Keywords: high gradient magnetic separation; capture selectivity; magnetic wire; matrix

1. Introduction

High gradient magnetic separation (HGMS) is an effective method for the concentration or removal of paramagnetic particles from various suspensions [1], and it has been widely applied in the field of mineral processing, for recovery of fine weakly magnetic minerals and for removal of such minerals from non-metallic ores [2,3]. This method is achieved through the use of numerous small magnetic wires in matrix, to generate a high magnetic field gradient at the vicinity of the wires in background magnetic field, and consequently produce a sufficiently powerful magnetic capture force to magnetic particles [4]. Therefore, the HGMS performance is inherently dependent on the capture characteristics of the wires [5]. But, as we know, in the most cases HGMS separators were basically limited in the roughing or scavenging process of weakly magnetic ores such as hematite and ilmenite [6,7], as they produced relatively low-grade concentrates. Obviously, the low-grade concentrates were resulted from the relatively low capture selectivity of magnetic wires to magnetic particles in the HGMS separators.

Theoretically, magnetic particles would be captured onto the wire due to the action of magnetic capture force, while non-magnetic particles would flow through the wire. However, in practice, some non-magnetic particles will be inevitably entrained into magnetic deposits in HGMS process, reducing the capture selectivity of the wire. The significant deviation of practical wire capture from the theoretical one has attracted much attention over the years, and many people have reported the capture selectivity of magnetic wire to particles in HGMS process. For instance, it has been early reported that the low separation selectivity of HGMS is mainly resulted from the non-looseness of slurry flowing around the wire, and thus pulsating HGMS technology was invented; in this new technology, the pulsating slurry improves the looseness of slurry, and significantly enhances the HGMS selectivity [8]. Ren et al. has comparatively investigated the capture selectivity of single wire and multi-wires, and found that the single wire achieved obviously superior selectivity to that of the multi-wires [9]. Moreover, some other researchers reported the dependence of capture selectivity on the shape of magnetic wire [10,11]. The build-up of magnetic particles onto magnetic wire, the size matching of magnetic wire to independent particles and the capture dynamics of magnetic wire in fluid were also reported, in order to understand the basic principles of wire magnetic capture in HGMS process [12–14].

These investigations are encouraging, as they contributed to improve HGMS performance. However, the inherent and detailed entraining mechanism of non-magnetic particles into magnetic deposits on a magnetic wire surface is insufficient, which has restricted the development of HGMS technologies. In this work, the selective capture of magnetic wire to particles in high gradient magnetic field was theoretically described and simulated using COMSOL Multiphysics. These theoretical and simulated results were then basically validated with the experimental magnetic capture to ilmenite ore, using an experimental magnetic capture method.

2. Theoretical

2.1. Magnetic Capture of Wire to Particles

As shown in Figure 1, suppose a cylindrical magnetic wire of diameter a is placed in the separating zone of a HGMS process and a uniform magnetic field B_0 is applied in the zone. A paramagnetic magnetic particle of volume V and magnetic susceptibility K is carried through the wire by slurry flowing at a velocity of v_0. It is noted that the slurry is parallel to the applied filed and they are perpendicular to the axis of wire.

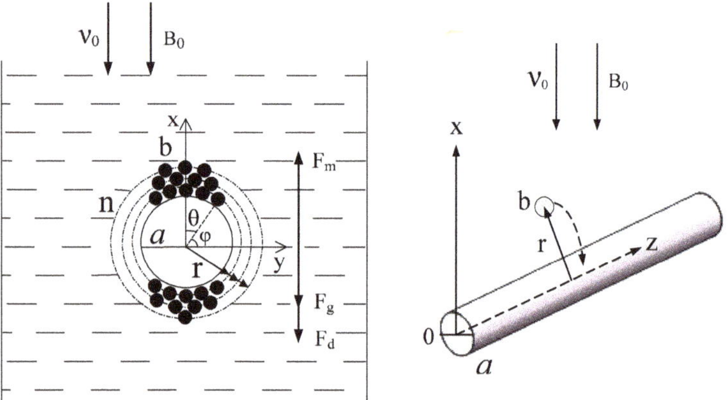

Figure 1. High gradient magnetic separation (HGMS) process (**left**) and magnetic capture of cylindrical wire to magnetic particle (**right**).

In the above process, the dominating forces that should be considered are magnetic force F_m, hydrodynamic drag F_d and gravity force F_g.

2.1.1. Magnetic Force

The magnetic force acting onto the particle is proportional to the magnetic induction and gradient, as shown in Equation (1).

$$F_m = \frac{1}{6\mu_0}\pi K b^3 B \, grad B,$$ (1)

where, μ_0 is the permeability of free space, b is the diameter of particle.

In the polar coordinate P (r, θ), Equation (1) is further written as:

$$\begin{cases} F_{mr} = \frac{\pi}{12\mu_0} K b^3 B_0^2 \frac{a^2}{r^3}(\frac{a^2}{4r^2} + cos2\theta) \\ F_{m\theta} = \frac{\pi}{12\mu_0} K b^3 B_0^2 \frac{a^2}{r^3} sin2\theta \end{cases},$$ (2)

where, r is the capture radius of magnetic force from the center of wire to that of particle.

2.1.2. Hydrodynamic Drag

The hydrodynamic drag is the main force against the capture of magnetic force, and it could be calculated using Stokes formula:

$$F_d = 3\pi\eta b(v_l - v_p),$$ (3)

where, v_l and v_p are the velocity of slurry and particles, respectively.

Suppose the slurry is inviscid and incompressible, and then the velocity of slurry flowing through the wire could be calculated using Equation (4):

$$\begin{cases} v_r = v_0 \cdot (1 - \frac{a^2}{4r^2})cos\theta \\ v_\theta = -v_0 \cdot (1 + \frac{a^2}{4r^2})sin\theta \end{cases},$$ (4)

Combing Equations (3) and (4), the hydrodynamic drag could be further written as:

$$\begin{cases} F_{dr} = 3\pi\eta b[v_0 \cdot (1 - \frac{a^2}{4r^2})cos\theta - \frac{dr}{dt}] \\ F_{d\theta} = 3\pi\eta b[-v_0 \cdot (1 + \frac{a^2}{4r^2})sin\theta - r\frac{dr}{d\theta}] \end{cases},$$ (5)

where, $\frac{dr}{dt}$ and $r\frac{d\theta}{dt}$ are the radial component and tangential components of particle velocity.

2.1.3. Gravitational Force

The effective gravitational force acting onto the particle is:

$$\begin{cases} F_{gr} = (\rho_p - \rho_f)gVcos\theta \\ F_{g\theta} = (\rho_p - \rho_f)gVsin\theta \end{cases},$$ (6)

where, ρ_p and ρ_f are the density of particle and fluid, respectively; g is the gravitational acceleration.

2.1.4. Required Magnetic Force for Effective Capture

For a particle to be effectively captured onto the wire, the interactive force between magnetic wire and particle should be attractive, i.e., magnetic force counteracts competing force:

$$\begin{cases} F_{mr} \geq F_{d\theta} + F_{gr} \\ F_{m\theta} \geq F_{d\theta} + F_{g\theta} \end{cases},$$ (7)

2.2. Motion Trajectory of Particle

2.2.1. Trajectory of Independent Particles

The motion of particles is determinable by Newton's second law:

$$\vec{F} = m\vec{a},\tag{8}$$

where, $\vec{F}$ is the resultant force acting onto the particle, m and $\vec{a}$ are the mass and accelerated velocity of particle, respectively.

For an independent particle, the interactions between the particle and slurry, the van der Waals force and the interactions between particles is negligible, so the trajectory of particle is mainly determined by the magnetic force, hydrodynamic drag and gravitational force:

$$\frac{d(m_p v)}{dt} = \vec{F}_d + \vec{F}_m + \vec{F}_g,\tag{9}$$

Combing Equations (3), (5), (6), and (9), the motion equation of the independent particle was obtained:

$$\begin{cases} \frac{d_r}{d_t} = \frac{2v_0}{a}\left(1 - \frac{a^2}{4r^2}\right)cos\theta - \frac{8KB_0^2 b^2}{9\eta\mu_0 a^2}\left(\frac{1}{r^5} + \frac{cos2\theta}{r^3}\right) \\ r\frac{d_\theta}{d_t} = -\frac{2v_0}{a}\left(1 + \frac{a^2}{4r^2}\right)sin\theta - \frac{8KB_0^2 b^2}{9\eta\mu_0 a^2} \cdot \frac{sin2\theta}{r^3} \end{cases},\tag{10}$$

Based on Equation (10) and the COMSOL Multiphysics simulation, the trajectories of independent particles are shown in Figure 2, and the related boundary conditions for the simulation are listed in Table 1. It is noted that in the simulation, only six magnetic particles and six non-magnetic particles were fed into the separating zone from the inlet at the initial time, so that the interactions between the particles could be neglected in the simulation; in addition, the interactions between the particles and slurry are not considered in the simulation.

Table 1. Boundary conditions for COMSOL Multiphysics simulation.

Boundary Conditions	Set Value
Relative magnetic permeability of slurry	1
Relative magnetic permeability of wire	7000
Background magnetic induction	0.6 and 1.0 T
Viscosity of slurry	1.2×10^{-3} Pa·s
Density of slurry	1.3×10^3 kg/m^3
Volume magnetic susceptibility of particle	2.57×10^{-3}
Density of particle	4.5×10^3 kg/m^3
Diameter of particle	0.10 mm
Feed velocity	0.04, 0.06 and 0.10 m/s

From Figure 2, the trajectories of magnetic particles at the initial time are similar to those of non-magnetic particles, as they are mainly dragged by the hydrodynamic drag during this period. However, when the particles move around the wire, the trajectories of magnetic particles are significantly different from those of non-magnetic particles. The magnetic particles would be attracted by the magnetic wire due to the action of magnetic force, resulting in their variation in their trajectories. But, when the non-magnetic particles move around the wire, they flow through the wire under the action of hydrodynamic drag, with two non-magnetic particles colliding with the wire due to their inertia effect under the particular condition of 1.2 T magnetic induction and 3.0 mm/s flow velocity; but, they were instantaneously bounced back by the magnetic wire.

It can be seen from the above discussion that under the ideal condition, the magnetic wire in the magnetic field only captures magnetic particles, and the non-magnetic particles would not be captured onto the magnetic wire as a result of the hydrodynamic drag.

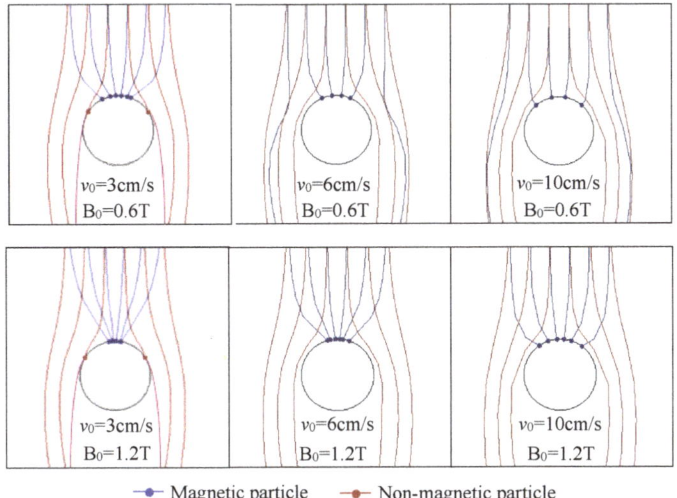

Figure 2. Simulated trajectories of independent particles in capture of single-wire.

2.2.2. Trajectory of Interactive Particles

In the actual HGMS process, the interactions between particles, and those between particles and slurry, have definitely their effects on the separation performance, due to the fact that the particles are fast and massively fed into the separating zone of HGMS separator. These interactions should be considered in the calculation and simulation for the motion trajectories of particles:

$$\frac{d(m_p v)}{dt} = \vec{F}_d + \vec{F}_m + \vec{F}_g + \vec{F}_p + \vec{F}_s, \tag{11}$$

where, $\vec{F}_p$ and $\vec{F}_s$ are the interactions between particles, and those between particles and slurry, respectively.

The interactions between the particles in Equation (11) could be written as:

$$\vec{F}_p = \frac{A}{3}\left[\frac{2b^2(x+2b)}{(x^2+4bx)^2} + \frac{2b^2}{(x+2b)^2} - \frac{x+2bx}{x+4b} + \frac{1}{x+2b}\right], \tag{12}$$

where, A is the constant of Hamaker, and x is the distance between adjacent particles.

Also, the interactions between particles and slurry could be directly calculated using the COMSOL Multiphysics 5.4.

Using the same boundary conditions (Table 1) as those for the simulation independent particles, the simulated trajectories of interactive particles are shown in Figure 3. It is noted that in the simulation the particles are continuously and massively fed into the separating zone.

Combing Figures 2 and 3, it is clear that the trajectories of interactive particles are significantly different from those of independent particles. As can be seen in Figure 3, most of non-magnetic particles directly flow through the wire, but a small part of the particles would be entrained into the magnetic deposit captured on the wire surface. These particles could not be released from the deposits under the action of interactions between adjacent particles, which results in the reduced capture selectivity of the wire. It was found from Figure 3 that under the higher magnetic induction and lower slurry velocity, the non-magnetic particles would be more easily entrained into the magnetic deposits on the wire surface.

In addition, the interactive particles were captured onto the wire in multi-layers, due to the fact that these particles were continuously and massively fed into the zone. In this multi-layer capture, the particles entrained into the inner layers would be wrapped by outer particles, leading to the difficulty for releasing the entrained particles. It is noted that the particles in multi-layers form a dense deposit on the wire surface under the action of strong magnetic force and the interactions between particles, preventing the entrained particles to be released from the deposits.

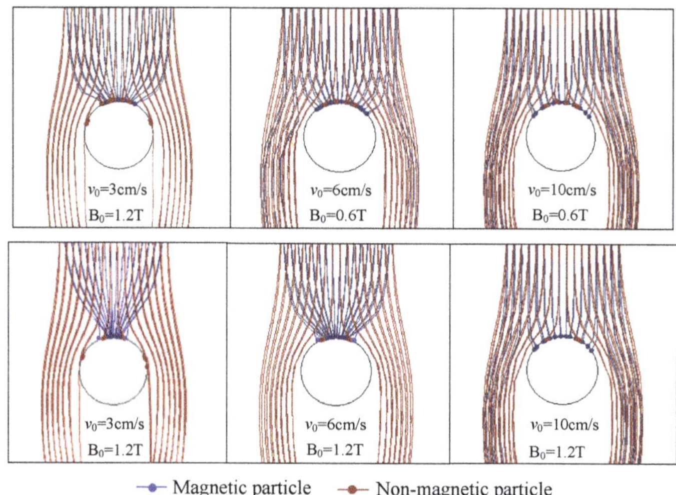

Figure 3. Simulated trajectories of interactive particles in capture of single-wire.

2.3. Fluid Characteristics around Single-Wire and Multi-Wires

In practice, matrix is made of numerous magnetic wires, so that a comparative investigation on the magnetic capture of single-wire and multi-wires would provide a crucial foundation for improving HGMS performance. It is noted that the difference in the capture characteristics of single-wire and multi-wires in the HGMS process is mainly resulted from their different fluid characteristics around the wires. Therefore, in this section, the fluid characteristics around the single-wire and multi-wires were respectively simulated using the COMSOL Multiphysics, as shown in Figure 4.

From Figure 4, for the single wire, there is a roundabout flow around the wire, and the slurry flows out of the wires, contributing to release the non-magnetic particles. However, after the particles are entrained in the upstream and downstream of the wire surface, the flow velocity of slurry on these two areas is nearly zero, increasing the difficulty for releasing the non-magnetic particles. For the multi-wires, the slurry flowing around the wire would be affected by adjacent wires, and it cannot flow through the wires freely. Comparing the fluid characteristics around the single-wire and multi-wires, the flowability of slurry in the multi-wires is much weaker than that through the single-wire, so that the capture selectivity of multi-wires gets lower than that of the single-wire.

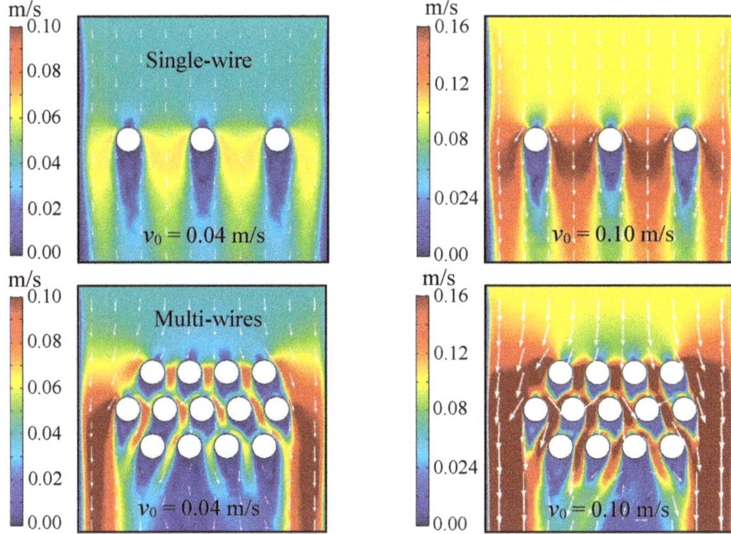

Figure 4. Fluid characteristics around single-wire (**above**) and multi-wires (**below**).

3. Experimental

3.1. Cyclic Pilot-Scale Pulsating HGMS Separator

A SLon-100 cyclic pilot-scale pulsating HGMS separator was used for the present investigation. This separator is fed periodically and its detailed description on the separation mechanism was early reported by Chen et al. [5]. When the separator was operated, via feed box the slurry was fed through the magnetic wire located in the separating zone of the separator, with magnetic particles captured onto the wires and non-magnetic particles flowing out of the zone. When a batch of feed was finished, the energizing current was switched off and the magnetic particles captured onto the wires were washed out to get a magnetic product.

3.2. Description of Material

A typical ilmenite ore assaying 13.04% TiO_2 was used for the capture investigation, and the volume magnetic susceptibility of ilmenite particles in the ore is 2.57×10^{-3}. The material is less than 0.15 mm and 56.73% of the material is smaller than 0.074 mm.

3.3. Method

Magnetic Capture Analysis (MCA) method [15] was used to investigate the capture selectivity of single wires. As shown in Figure 5, in the method circular holes were regularly drilled in two non-magnetic plates, which were used for inserting magnetic wires. In the investigations, the wires were inserted in the plates at a sufficiently large spacing of 6 mm to avoid the magnetic coupling effect between the wires.

In this investigation, 3 wires were used in the MCA method, and the material was fed into the separating zone at two different feed modes, to achieve the magnetic capture of wire to particles under ideal and actual conditions, as shown in Figure 6. In the actual condition, 40 g material was fast fed into the separating zone within 10 s, and was maintained in the zone for a sufficiently long capturing time of 4 min. In the ideal condition, 40 g material was uniformly divided into 8 fractions, and each fraction was slowly fed into the separating zone within 10 s and maintained for 4 min; after that, another fraction was fed after 10 s interval, until all material was fed.

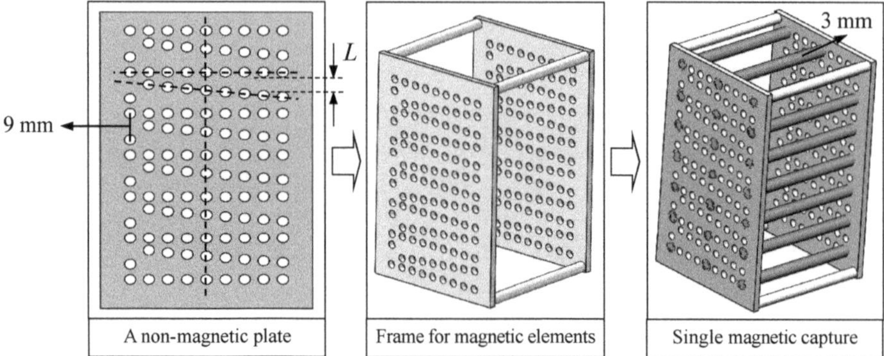

Figure 5. Schematic diagram on operational principle of Magnetic Capture Analysis (MCA) method.

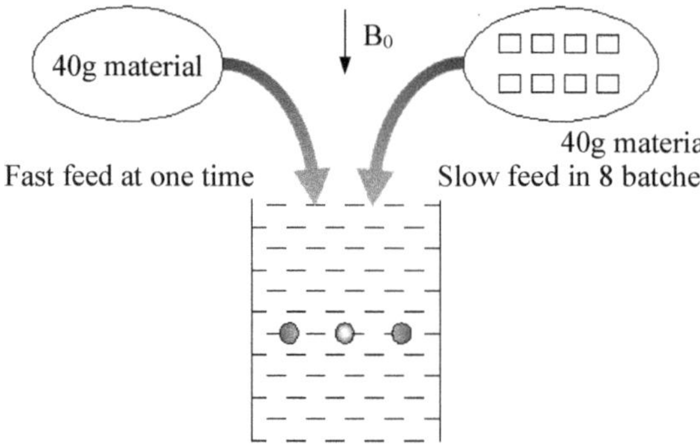

Figure 6. Fast (actual condition) and slow (ideal condition) feed models for material into separating zone.

The capture selectivity of multi-wires was investigated using a real matrix, under the exact same operating conditions as those of single wires. But for this investigation, 200 g of material was fast fed into the separating zone within 30 s, and the material was maintained in the zone for 8 min.

For all the investigations, 2- and 3-mm diameter magnetic wires were used, and the slurry flows through the separating zone at a velocity of 4.0 cm/s. In the investigations, the TiO_2 grade of capture products and their mass weight for unit length of wire (g/cm) were respectively used to evaluate the capture selectivity and capture capability of wires.

4. Results and Discussion

4.1. Effect of Magnetic Induction on Capture Selectivity of Single-Wire

Controlling the pulsating frequency of slurry at 180 r/min, i.e., slurry moving up and down 180 times per minute in the separating zone, and the particle size of 76.56% below 0.074 mm, the magnetic capture selectivity of single wires, under the slow and fast feed modes as shown in Figure 6, was comparatively investigated, with magnetic induction increasing from 0.4 T to 1.4 T. From Figure 7, for both the two feed-modes, the TiO_2 grade of captured ilmenite products was decreased with increase in the magnetic induction. However, it is obvious that the decrease in the TiO_2 grades from the fast feed are much more significantly than that from the slow feed, particularly after 0.8 T.

The most important discovery in this investigation is that the wires in the slow feed mode have achieved higher capture selectivity than that in the fast feed mode, and the experimental results are consistent with the theoretical analyses afore. For instance, at the magnetic induction of 0.8 T, the TiO$_2$ grade of capture product from the 3 mm wire under the slow feed condition reached as high as 36.78%, while it was only 28.32% for the fast feed. According to the theoretical discussions, in the actual condition (fast feed) non-magnetic particles would be easily entrained into the captured deposits under the action of interactions between particles, so that the capture selectivity of the wire in the actual condition is much lower than that from the ideal condition (slow feed). In addition, in the actual condition, the particles were captured on the wire surface in multi-layers, due to the fact that the particles are fast and massively fed into the separating zone. These multi-layer particles would form a dense deposit on the wire surface under the action of strong magnetic force and the interactions between particles, preventing the entrained particles to be released from the deposit and subsequently reducing the capture selectivity of the wire.

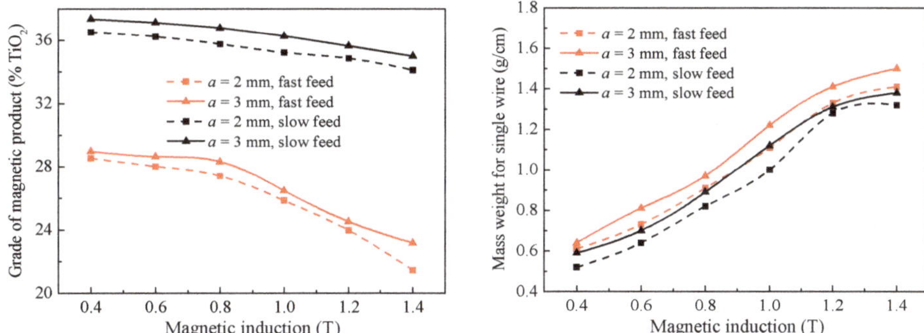

Figure 7. Effect of magnetic induction on capture selectivity (**left**) and capture capacity (**right**) of single wire with different feed modes.

It was found that the selective difference between the actual and ideal conditions increased with increase in the magnetic induction. Such an increase in the difference may be mainly due to the fact that under the high magnetic induction, non-magnetic particles more easily collided with the wire in the actual condition, as shown in Figure 3, and so it is much more difficult for the entrained particles to release from the magnetic deposits in the higher induction.

Also, the mass weight of capture product increased with increase in the magnetic induction, and the wires in the ideal condition have achieved slightly higher capture mass weight than that from the wires in the actual condition. It should be noted that this investigation aims to investigate the capture selectivity of wires in high gradient magnetic field, and therefore only the TiO$_2$ grade of capture product was considered in the following investigations.

4.2. Effect of Pulsating Frequency on Capture Selectivity of Single-Wire

It is well known that the pulsation of slurry produces a relaxed effect to particles in the slurry, which improves the capture selectivity of magnetic wires. As shown in Figure 8, similar to the effect of magnetic induction as discussed above, the wires from the slow and fast feed models have basically the same capture trend while the pulsating frequency were gradually increased; but again, the effect of pulsating frequency on the capture selectivity of wires from the slow feed is much more gentle than that from the wires in fast feed. It was also found that the wires in the slow feed mode achieved a higher capture selectivity than that from the wires of fast feed mode.

As discussed afore, the entrainment of non-magnetic particles in the magnetic deposits may be mainly resulted from the interactions between particles, and the hydrodynamic force is the main force competing the interaction in the capture process. As the pulsation of slurry increases, the hydrodynamic force acting onto the particles weakens the interaction and facilitates the release of non-magnetic particles from magnetic deposits, thereby improving the capture selectivity of wires. But in the slow feed, non-magnetic particles are rarely entrained into the captured deposits and the magnetic wires achieved a high capture selectivity, so that the capture selectivity of wires in the slow feed is not so sensitive to the pulsating frequency.

However, it is noted that the entrainment of non-magnetic particles could not be completely eliminated in the pulsating slurry. From Figure 8, even at high pulsating frequency, the TiO_2 grade of magnetic product from the fast feed mode is still lower than that from the slow one.

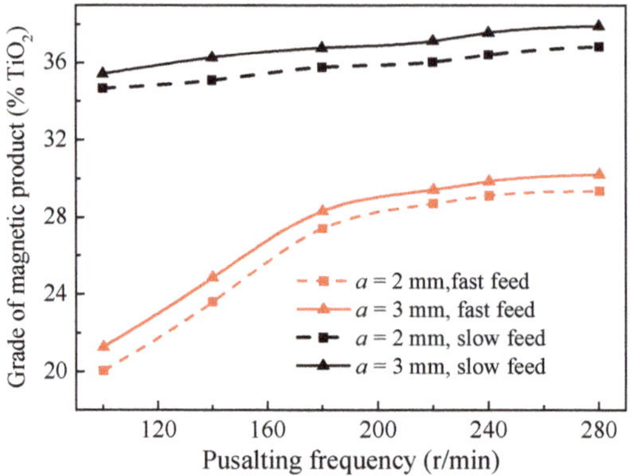

Figure 8. Effect of pulsating frequency on capture selectivity of single-wire with different feed modes.

4.3. Effect of Particle Size on Capture Selectivity of Single-Wire

It is well known that the particle size determines the liberation degree of material and so that produces significant effect on the capture selectivity of magnetic wire, as shown in Figure 9. It should be noted that this effect is investigated under magnetic induction of 0.8 T and pulsating frequency of 180 r/min. From Figure 9, for both the two feed-modes, the TiO_2 grades of captured deposits were increased with decrease in the particle size of feed material from 56.73% to 76.56% below 0.074 mm, and they approached the maximum values with further decrease in the particle size to 91.23% below 0.074 mm. It was also found that the wires in the slow feed mode achieved a higher TiO_2 grade than that from the wires in fast feed mode, and the grade difference between the fast and slow feeds increased with decrease in the particle size.

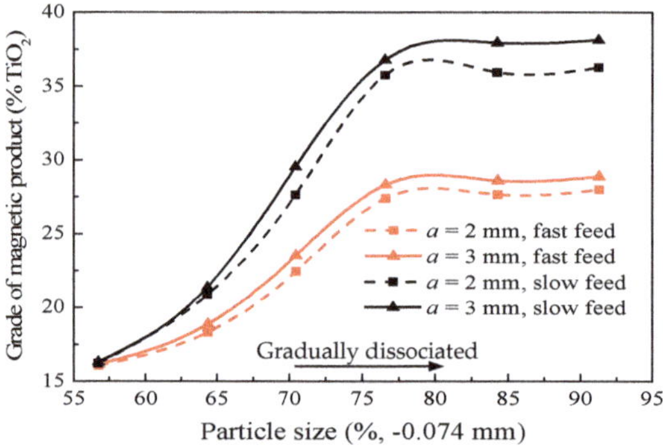

Figure 9. Effect of particle size on capture selectivity of single wire with different feed modes.

For large particle size, the material was not fully dissociated, and then many intergrowth minerals combining magnetic and non-magnetic components may be captured on the wire surface, reducing the TiO_2 grade of captured deposits. With decrease in the particle size, the intergrowths would be gradually liberated, which reduces the number of intergrowth minerals in the magnetic products and improves their TiO$_2$ grades. However, in the fast feed mode, many non-magnetic particles would be easily entrained into the magnetic deposits, as clearly described above, so that with decreasing particle size, the increase in the TiO$_2$ grades of magnetic products in the slow feed is not so significant as that in the fast feed.

4.4. Comparation of Capture Selectivities between Single Wire and Multi-Wires

As mentioned afore, the multi-wires have different capture characteristics from the single wire. Thus, in this section, the capture selectivity of single wire and multi-wires were comparatively investigated, with magnetic induction increasing from 0.4 T to 1.4 T, as shown in Figure 10. It is noted that the material was fed in the fast feed mode as mentioned afore. From Figure 10, the single wire and multi-wires produced two similar sets of capture selectivity, and the TiO$_2$ grades of magnetic products reduced with increase in the magnetic induction. However, it was found that the multi-wires produced much lower TiO$_2$ grades of magnetic products than those from the single wire, and the capture selectivity of multi-wires was much more sensitive to variations in the magnetic induction.

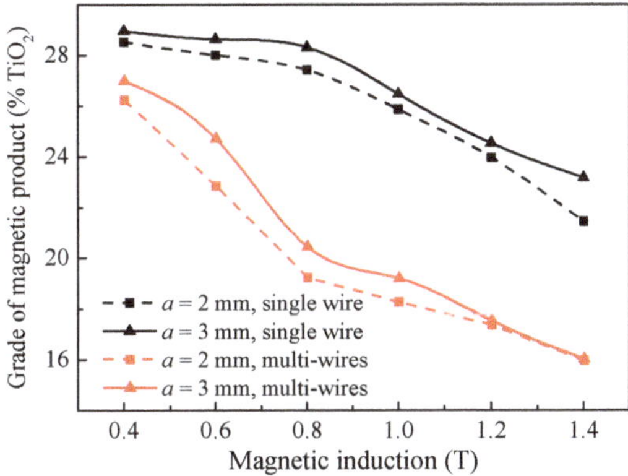

Figure 10. Capture selectivity of single wire and multi-wires.

The reduced capture selectivity of the multi-wires is mainly resulted from the poor flowability of slurry in the multi-wires, as shown in Figure 4. In addition, in the multi-wires, after an entrained particle is released from magnetic deposits on the wire surface, the particle may collide with other adjacent wires, increasing the probability of non-magnetic particle entraining into capture deposits. Therefore, the separating selectivity in the actual HGMS process with multi-wires is much lower than that from the ideal one.

5. Conclusions

1. The motion trajectories of a few of particles in the ideal HGMS condition is significantly different from those of a lot of particles in the actual HGMS condition, as the interactions between particles and those between particles and slurry have their significant effects on the trajectories. In addition, the captured particles form a compact deposit on the wire surface due to the interactions between the particles, preventing entrained particles to be released from the deposit; therefore, non-magnetic particles would be more easily entrained into the deposit in the actual HGMS condition than those in the ideal HGMS condition. The flowability of slurry in the multi-wires is much weaker than that through a single-wire.

2. The experimental results are consistent with the theoretical analyses. The wires in slow feed mode produced a higher capture selectivity than that in fast feed mode, and their difference in the selective capture increased with increase in the magnetic induction and decrease in the pulsating

frequency. The multi-wires produced much lower TiO_2 grades of magnetic products than those from the single wire, as they produce different effects on the flowability of slurry.

3. The selective capture of magnetic wire to magnetic particles could be improved in a pulsating slurry, but the entrainment of non-magnetic particles in the magnetic deposits on wire surface could not be fully eliminated.

This investigation provides a beneficial understanding on the selective capture of magnetic wires to magnetic particles in a HGMS process.

Author Contributions: Methodology, J.Z. and L.C.; validation, J.Z. and F.Y.; formal analysis, J.Z.; investigation, J.Z.; resources, L.C. and X.T.; data curation, J.Z. and F.Y.; writing—original draft preparation, J.Z.; writing—review and editing, J.Z. and L.C.; supervision, L.C. and X.T.; project administration, L.C. and X.T.; funding acquisition, L.C. and X.T.

Funding: This research was funded by National Natural Science Foundations of China (Grant No. 51874152) and the Key Program for Applied Basic Research of Yunnan Province (Grant No. 2016FA051).

Conflicts of Interest: The authors declare no conflict of interest.

References

1. Padmanabhan, N.; Sreenivas, T. Process parametric study for the recovery of very fine size uranium values on super-conducting high gradient magnetic separator. *Adv. Powder Technol.* **2011**, *22*, 131–137. [CrossRef]
2. Chen, L.; Xiong, D.; Huang, H. Pulsating high-gradient magnetic separation of fine hematite from tailings. *Miner. Metall. Proce.* **2009**, *26*, 163–168. [CrossRef]
3. Chen, L.; Liao, G.; Qian, Z.; Chen, J. Vibrating high gradient magnetic separation for purification of iron impurities under dry condition. *Int. J. Miner. Process.* **2012**, *130*, 136–140. [CrossRef]
4. Zheng, X.; Wang, Y.; Lu, D. Study on capture radius and efficiency of fine weakly magnetic minerals in high gradient magnetic field. *Miner. Eng.* **2015**, *74*, 79–85. [CrossRef]
5. Chen, L.; Liu, W.; Zeng, J.; Ren, P. Quantitative investigation on magnetic capture of single wires in pulsating HGMS. *Powder Technol.* **2017**, *313*, 54–59. [CrossRef]
6. Xiong, D. SLon magnetic separator applied in the ilmenite processing industry. *Phys. Sep. Sci. Eng.* **2004**, *13*, 119–126.
7. Zeng, W.; Xiong, D. The latest application of SLon vertical ring and pulsating high-gradient magnetic separator. *Miner. Eng.* **2003**, *16*, 563–565. [CrossRef]
8. Xiong, D.; Liu, S.; Chen, J. New technology of pulsating high gradient magnetic separation. *Int. J. Miner. Process.* **1998**, *54*, 111–127. [CrossRef]
9. Ren, P.; Chen, L.; Liu, W.; Shao, Y.; Zeng, J. Comparative investigation on magnetic capture selectivity between single wires and a real matrix. *Results Phys.* **2018**, *8*, 180–183. [CrossRef]
10. Li, W.; Han, Y.; Xu, R.; Gong, E. A preliminary investigation into separating performance and magnetic field characteristic analysis based on a novel matrix. *Minerals* **2018**, *8*, 94. [CrossRef]
11. Zheng, X.; Wang, Y.; Lu, D. Particle capture efficiency of elliptic cylinder matrices for high-gradient magnetic separation. *Sep. Sci. Technol.* **2016**, *51*, 2090–2097. [CrossRef]
12. Zeng, J.; Tong, X.; Ren, P.; Chen, L. Theoretical description on size matching for magnetic element to independent particle in high gradient magnetic separation. *Miner. Eng.* **2019**, *135*, 74–82. [CrossRef]
13. Zheng, X.; Wang, Y.; Lu, D. Study on buildup of fine weakly magnetic minerals on matrices in high gradient magnetic separation. *Physicochem. Probl. Miner. Process.* **2017**, *53*, 94–109.
14. Svoboda, J. A realistic description of the process of high-gradient magnetic separation. *Miner. Eng.* **2001**, *14*, 1493–1503. [CrossRef]
15. Lu-Zheng, C.; Wen-Bo, L.; Qing-Fei, X.; Jian-Wu, Z.; Xiong, T.; Zheng, Y.-M. An Experimental Method for Magnetic Capture Analysis of Single Magnetic Wires. China Patent No. 2016 1057 4053.3, 21 June 2016. (In Chinese).

Technical Note

Separation of Monazite from Placer Deposit by Magnetic Separation

Kwanho Kim and Soobok Jeong *

DMR Convergence Research Center, Korea Institute of Geoscience and Mineral Resources, Daejeon 34132, Korea; khkim@kigam.re.kr
* Correspondence: sbjeong@kigam.re.kr; Tel.: +82-42-868-3576

Received: 23 January 2019; Accepted: 19 February 2019; Published: 28 February 2019

Abstract: In this study, mineralogical analysis and beneficiation experiments were conducted using a placer deposit of North Korea, on which limited information was available, to confirm the feasibility of development. Rare earth elements (REEs) have vital applications in modern technology and are growing in importance in the fourth industrial revolution. However, the price of REEs is unstable due to the imbalance between supply and demand, and tremendous efforts are being made to produce REEs sustainably. One of them is the evaluation of new rare earth mines and the verification of their feasibility. As a result of a mineralogical analysis, in this placer deposit, monazite and some amount of xenotime were the main REE-bearing minerals. Besides these minerals, ilmenite and zircon were the target minerals to be concentrated. Using a magnetic separation method at various magnetic intensities, paramagnetic minerals, ilmenite (0.8 T magnetic product), and monazite/xenotime (1.0–1.4 T magnetic product) were recovered selectively. Using a magnetic separation result, the beneficiation process was conducted with additional gravity separation for zircon to produce a valuable mineral concentrate. The process resulted in three kinds of mineral concentrates (ilmenite, REE-bearing mineral, and zircon). The content of ilmenite increased from 32.5% to 90.9%, and the total rare earth oxide (TREO) (%) of the REE-bearing mineral concentrates reached 45.0%. The zircon concentrate, a by-product of this process, had a Zr grade of 42.8%. Consequently, it was possible to produce concentrates by combining relatively simple separation processes compared to the conventional process for rare earth placer deposit and confirmed the possibility of mine development.

Keywords: rare earth elements; monazite; placer deposit; beneficiation; ilmenite; magnetic separation

---

1. Introduction

Rare earth elements (REEs) are a group of elements belonging to the lanthanide series, which ranges from lanthanum to lutetium; the group also contains scandium and yttrium with similar chemical properties [1,2]. Due to their unique properties, REEs are widely used in applications such as magnets, battery alloys, and metal alloys. The importance of REEs is growing due to their increasing application in modern technology and their consequent role in the fourth industrial revolution.

However, despite this increasing demand, the supply of REEs is not stable due to regionally biased production. According to the United States Geological Survey on REE production, more than 80% of the world's supply since 1998 has come from China, and this ratio increased to 95% in the mid-2000s [3]. In 2009, the export quota and tax restrictions imposed by the Chinese government resulted in an imbalance in the demand and supply, leading to a dramatic increase in REE prices. In response to rising prices, several companies in the United States, Australia, Brazil, and Russia began to produce REEs. Worldwide REE prices have almost returned to normal after five years because of excessive REE supply and the abolition of Chinese restrictions in 2015. Many new companies have shut down operations due to economic problems. As a result, China's influence on the supply of REEs will likely increase, increasing the probability of the return of the aforementioned instability in production.

REEs do not exist as natural metals and are contained in various minerals in a substituted state. However, out of over 250 REE-bearing minerals, only three (bastnasite, monazite, and xenotime) are currently produced on a commercial scale. Bastnasite is the main REE-bearing mineral in large mines (Mountain Pass, CA, USA, and Bayan Obo, China), and the other two exist as heavy mineral sand deposits [4]. To develop beneficiation flowsheets of various REE-bearing deposits, extensive research has been devoted by many researchers [5–11]. In particular, the beneficiation process for placer deposits is well established and includes a combination of gravity, magnetic, electrostatic, and, at times, flotation processes [12–14].

Apart from the well-established beneficiation process for placer deposits, the application of the unit process should be modified according to the mineralogy of the sample. In this study, a placer deposit from North Korea was used as a feed sample. However, the mineralogy data and beneficiation characteristics of the North Korean placer deposit sample are not well known. Therefore, in this study, the mineralogy of a placer deposit in North Korea was investigated and a beneficiation process, especially magnetic separation, was applied to separate valuable mineral selectively and examine the feasibility of resource development project.

2. Materials and Methods

2.1. Materials

The feed sample used in this study was obtained from a placer deposit of the Sam-Cheon area in North Korea. According to geological data available on North Korea, there are two REE mines, namely Wolbong and Ryeonsan, where the ore body comprises sandy-gravel layers in alluvium [15]. Figure 1 shows the geological map of Rimjingang belt in the middle Korean Peninsula and the location of two REE mines and Sam-Cheon area [16]. As shown in Figure 1, the samples were collected from the bottom of the active river channel and sand bar in the Sam-Cheon area near two REE mines. It was confirmed that in this mining site, the gravity separation methods using a spiral concentrator and shaking table were preliminary conducted to concentrate heavy minerals in the placer deposit. Therefore, in this study, the preliminary concentrated heavy mineral sample was used as the feed sample for the beneficiation process.

A total of 200 kg of feed sample was well mixed and divided into four portions, and one of four portions was selected and further divided into four portions repeatedly. The representative sample for sample analyses was obtained by repeating this procedure. Many analytical instruments were used to identify the feed sample before and after the experiment. X-ray diffraction (XRD; X'pert MPD, Philips, Malvern, UK) was used for mineral constituent analysis; inductively coupled plasma optical emission spectroscopy (ICP-OES; Optima-5300DV, Perkin Elmer, Waltham, MA, USA), inductively coupled plasma mass spectroscopy (ICP-MS; Elan DRC-II, Perkin Elmer, Waltham, MA, USA), and X-ray fluorescence spectrometer (XRF; MXF-2400, Shimadzu, Kyoto, Japan) were used for the chemical composition analysis; and the mineral liberation analysis (MLA; MLA650F, FEI, Hillsboro, OR, USA) was used to evaluate the degree of liberation and mineral constituents.

For ICP-OES and ICP-MS analyses, a sample pretreatment method was basically carried out according to the USGS method for analyzing rare earth elements by ICP-MS. 0.1 g of sample put into carbon crucible with 0.6 g of Sodium peroxide (Na_2O_2; Sigma–Aldrich) The carbon crucible heated at muffle furnace about 550 °C for 30 min. After cooling the carbon crucible, 10 ml of 25% nitric acid (HNO_3; 70%, Sigma–Aldrich) was added to dissolve elements for 15 min. The remaining 25% nitric acid was evaporated, and then 1% nitric acid and deionized water (DI) were added to make a 100 mL solution. This solution was diluted 10 times or 1000 times according to the concentration of elements to be measured.

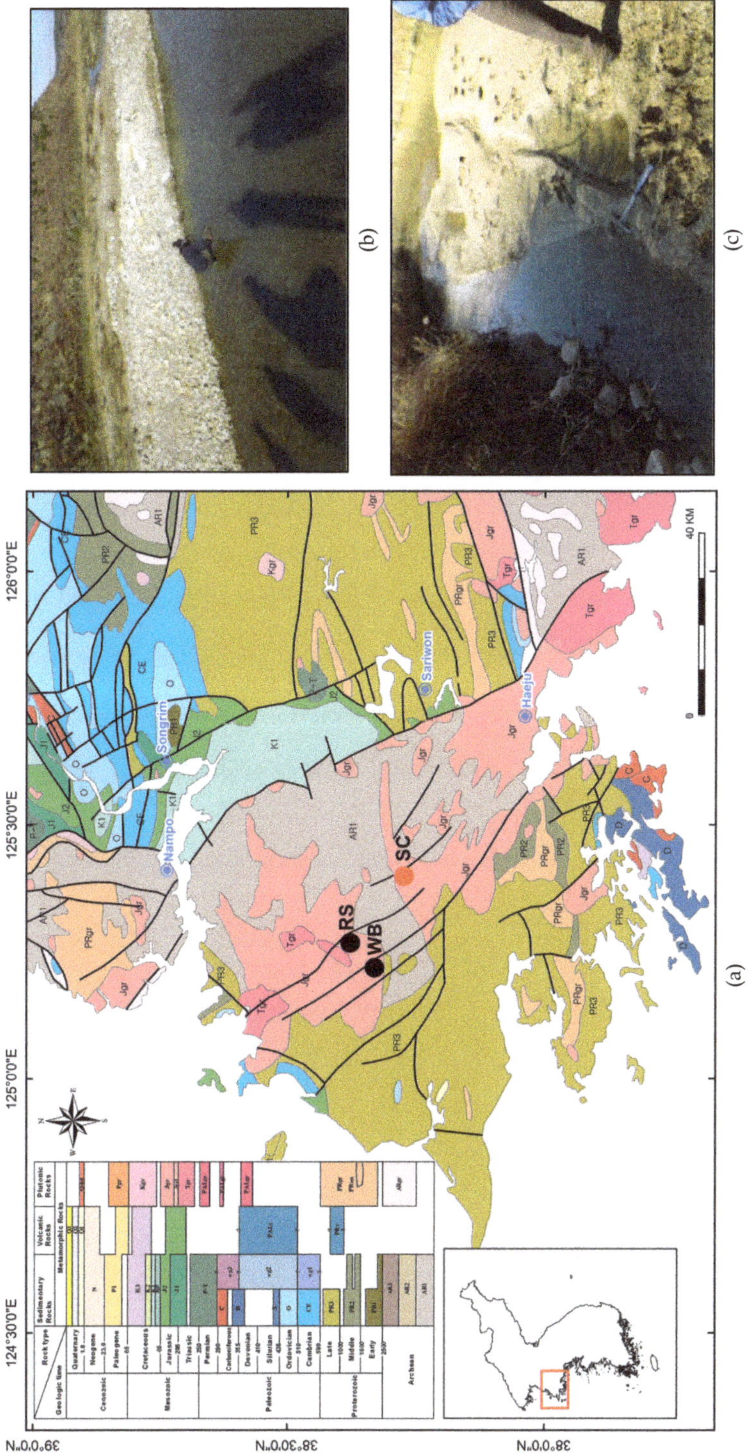

Figure 1. The geological map of the Sam–Cheon area (**a**) [16] and photographs of sampling location of feed sample (**b,c**). WB—Wolbong mine; RS—Ryeonsan mine; SC—Sam-Cheon area.

2.2. Methods

A laboratory-scale cross-belt magnetic separator (CBMS; Model EE112, Eriez, Erie, PA, USA) that could modulate the applied magnetic field by changing the current supply to the separator was employed to recover iron oxide and REE-bearing minerals. The sample was fed at approximately 200 g/min for 20 min using a vibrating feeder. The moving velocity of the feed carry conveyor was about 7.3 cm/s, and that of the cross-belt conveyor to recover magnetic minerals was about 31.5 cm/s. The applied magnetic intensity was increased from 0.4 T to 1.4 T. After recovering magnetic products at 0.4 T, the remaining non-magnetic sample was fed into a magnetic separator that was adjusted to a 0.2 T higher magnetic intensity for further separation. The magnetic separation tests were carried out sequentially in this way six times until the magnetic intensity reached 1.4 T.

The batch test was conducted using 10 kg of feed sample. In this process, besides the target minerals, a gravity separation process for recovering zircon as a by-product was added. To recover high-density zircon from non-magnetic product, a shaking table (No.13 Wilfley table, Humphreys, Jacksonville, FL, USA) was used. The operating conditions were adjusted based on the general conditions. The angle of the shaking table, shaking amplitude, and water flowrate were varied from 0.5° to 4°, from 10 to 20 mm, and from 5 to 15 L/min, respectively.

3. Results and Discussion

3.1. Feed Sample Analysis

The XRD pattern profiles of the feed sample in Figure 2 revealed the main constituent minerals, and mineral liberation analysis (MLA) results in Table 1 showed major and minor constituent minerals that had not been detected by XRD analysis owing to their low content. As a result, monazite was the main REE-bearing mineral and ilmenite was the main iron oxide mineral. Besides these two main minerals, quartz and some aluminosilicate minerals, which contain alumina (Al_2O_3) and silica (SiO_2), such as orthoclase, pyroxene, tourmaline, and muscovite, were detected as gangue minerals. Some metallic minerals such as garnet group minerals, zircon, rutile, and titanite had been identified in addition to main constituent minerals. However, their content was very low compared to those of monazite and ilmenite. Therefore, mainly monazite and ilmenite were considered as target minerals for the beneficiation process, whereas quartz was considered as a main gangue mineral from the feed sample.

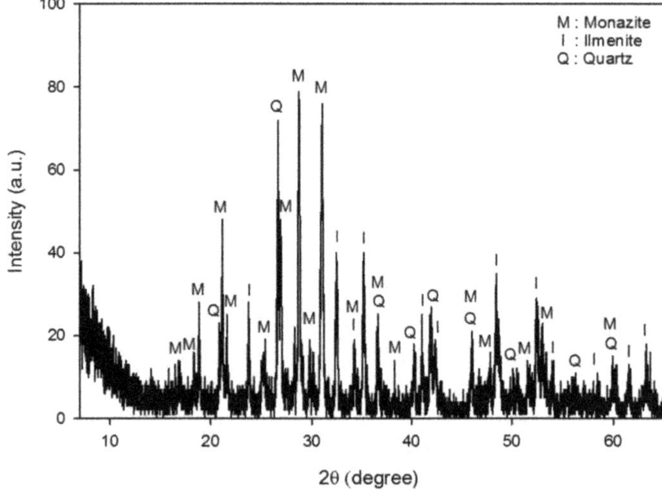

Figure 2. X-ray diffraction (XRD) patterns of the feed sample.

Table 1. Feed sample mineralogy (wt %) analyzed by mineral liberation analysis (MLA).

Mineral	wt %	Mineral	wt %
Monazite	42.6	Rutile	0.8
Ilmenite	33.8	Xenotime	0.6
Quartz	6.7	Tourmaline	0.4
[1] Garnet Group	6.3	Titanite	0.4
Orthoclase	2.9	Muscovite	0.4
Pyroxene	1.8	Magnetite	0.4
Zircon	1.7	Others	1.2

[1] Garnet Group: Almandine, Grossular garnet.

Table 2 shows the mass fraction (%) and concentration (%) of the main elements as a function of particle size in the feed sample. Basically, since the feed sample came from a placer deposit, it shows a relatively narrow particle size distribution (D_{10} = 130 µm, D_{50} = 220 µm, D_{90} = 400 µm), and more than 93% of the particles were distributed in the particle size range of 100–500 µm. The concentration of Fe, Ti, and TREO of the entire sample were 12.5%, 11.0%, and 20.5%, respectively. The concentration of Fe and Ti were high in coarse size fraction of 300 µm or more, and the one of TREO was high in intermediate size range between 100–300 µm. However, even though there was a difference in concentration as a function of particle size, selective separation of constituent minerals through simple particle size classification was not enough.

Table 2. Mass fraction (wt %) and concentration of main elements (%) of feed sample as a function of size fractions. TREO, Total rare earth oxide.

Size Fraction	Mass Fraction (wt %)	Concentration (%)		
		Fe	Ti	TREO
>500 µm	2.5	22.2	17.9	2.9
300–500 µm	22.6	15.1	13.2	13.2
212–300 µm	27.9	12.8	11.2	22.1
150–212 µm	31.6	10.9	9.6	24.6
106–150 µm	11.2	10.0	9.1	25.1
74–106 µm	2.4	10.8	9.7	19.5
<74 µm	1.9	10.2	8.0	12.7
Total	100.0	12.5	11.0	20.5

In the beneficiation process, it is important to understand and use the physical properties of the constituent minerals in order to recover the target minerals selectively. The most representative difference in physical properties of monazite, ilmenite, and quartz are the magnetic susceptibility and the specific gravity. Ilmenite and monazite, which are paramagnetic minerals, could be separated from quartz, which is a diamagnetic mineral, through magnetic separation. In addition, it has been reported in previous research [2,9] that a separation of ilmenite and monazite can be achieved by controlling the magnetic intensity. On the other hand, quartz with a relatively low density (2.65 g/cm^3) can be easily separated from ilmenite (4.8 g/cm^3) and monazite (4.8–5.5 g/cm^3), which have a high density, by gravity separation. However, it is difficult to separate ilmenite and monazite from each other owing to a similar specific gravity. Therefore, the magnetic separation method could be regarded as an effective method to separate ilmenite, monazite, and quartz selectively.

The liberation characteristics and particle associations between minerals in the feed sample were determined by MLA. Because the feed sample came from a placer deposit, the degree of liberation was expected to be high compared with a sample from open-pit or underground mining. The part of the polished section image of the feed sample obtained from MLA is shown in Figure 3a. The degree of liberation of each particle was found to be high. The particles shown in Figure 3a are mainly composed of "green" monazite, "brown" ilmenite, and "grey" quartz. To identify the degree of liberation of major minerals, the liberation characteristics of the three main minerals calculated in terms of mineral composition (wt %) are categorized into two groups based on a previous study [8]. The "liberated" category includes particles that have a mineral composition of more than 80%, and the "non-liberated" category includes those with mineral composition of below 80%. As can be seen in Figure 3b, the weight fractions of the minerals regarded as "liberated" particles exceeded 90% for all three minerals. The ratio of the "liberated" particles of monazite, in particular, was as high at 99.8%.

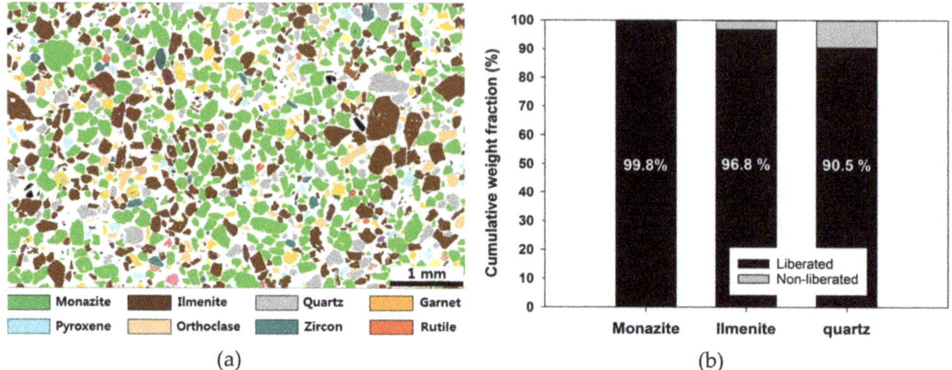

Figure 3. The part of polished section image (**a**) of feed sample analyzed by mineral liberation analysis (MLA) and degree of liberation (**b**) of main minerals in the feed sample.

Because the main valuable minerals of the sample had high densities (4.0–6.0 g/cm^3), gravity separation can be an effective method to remove light-density gangue minerals, such as quartz (density = 2.65 g/cm^3) and orthoclase (density = 2.55 g/cm^3). Therefore, a sink-float test was conducted using tetrabromoethane (density = 2.98 g/cm^3), as shown in Table 3. The results indicated that the yield of heavy minerals was approximately 86%, whereas that of light minerals was about 14%, and the contents of Fe, Ti, and TREO (%) were mainly distributed in the heavy mineral sample. Although the effect of the gravity separation method to recover target minerals was good, additional gravity separation was not necessary, since more than 86% of the feed sample was composed of heavy minerals. Therefore, the feed sample was directly subjected to the magnetic separation process without pretreatment.

Table 3. Sink-float experiment results of feed sample.

		Chemical Composition					
	Yield (wt %)	Fe		Ti		[1] TREO	
		[2] C (%)	[3] R (%)	C (%)	R (%)	C (%)	R (%)
Sink fraction (Heavy mineral)	86.2	14.4	97.2	12.7	99.5	23.7	99.7
Float fraction (Light mineral)	13.8	1.0	2.8	0.2	0.5	0.17	0.3

[1] TREO—Total rare earth oxide; [2] C—Grade; [3] R—Recovery.

3.2. Dry Magnetic Separation

Table 4 shows the dry magnetic separation results as a function of magnetic intensity. As mentioned in Section 2.2, the magnetic separation tests were conducted sequentially by increasing the magnetic intensity while recovering magnetic products at low magnetic intensity.

As a result, it could be confirmed that the content of Fe and Ti were concentrated at the magnetic intensity of 1.0 T or less. Because Fe and Ti mainly originate from ilmenite in feed sample, the grade of ilmenite ($FeTiO_3$) was estimated using the theoretical ratio of Fe/Ti (1.17) as follows:

$$\text{Grade of ilmenite (\%)} = \begin{cases} \dfrac{\text{Ti content } \{\%}{\text{theoretical ratio of Ti}/FeTiO_3} & if,\ Fe/Ti\ > 1.17 \\[2ex] \dfrac{\text{Fe content } \{\%}{\text{theoretical ratio of Fe}/FeTiO_3} & if,\ Fe/Ti\ < 1.17 \end{cases} \tag{1}$$

If the Fe/Ti ratio of a sample is larger than 1.17, indicating that excess Fe comes from other minerals, such as magnetite (Fe_3O_4), the grade of ilmenite is calculated on the basis of the Ti content by using theoretical ratio of Ti/$FeTiO_3$ (0.3156). On the other hand, if the Fe/Ti ratio of a sample is smaller than 1.17, indicating that excess Ti comes from other minerals, such as rutile (TiO_2), the amount of ilmenite is calculated on the basis of the Fe content by using theoretical ratio of Fe/$FeTiO_3$ (0.3681). As a result, at magnetic intensities of 0.6 T and 0.8 T, ilmenite was mainly recovered, and its grade was almost 90%. As magnetic intensity increased, the ilmenite grade decreased sharply and became only 1% at 1.2 T or higher magnetic intensity. Therefore, the recovery of ilmenite up to 0.8 T magnetic intensity was 94.3%, and most ilmenite was concentrated below 0.8 T.

Table 4. The grade and recovery of main minerals at various magnetic intensity products.

Magnetic Intensity (T)	Yield (wt %)	Ilmenite				REE				SiO$_2$		
		Fe (%)	Ti (%)	[1] C (%)	[2] R (%)	[3] LREO (%)	[4] HREO (%)	TREO (%)	[2] R (%)	[1] C (%)	[2] R (%)
Non-Magnetic	19.1	0.7	1.2	2.0	1.2	0.3	0.04	0.34	0.3	62.2	75.9
0.4	1.5	42.3	18.2	57.7	2.7	0.3	0.02	0.32	0.02	4.9	0.5
0.6	22.0	33.6	30.7	91.3	61.8	0.04	0.0	0.04	0.04	5.0	7.0
0.8	11.1	32.1	29.4	87.2	29.8	0.09	0.03	0.12	0.06	4.7	3.3
1.0	3.4	15.9	9.2	29.2	3.1	3.0	9.1	12.1	2.0	4.2	0.9
1.2	38.1	0.5	0.4	1.3	1.5	46.2	1.8	48.0	89.2	4.6	11.2
1.4	4.7	0.6	0.3	1.0	0.1	34.8	1.1	35.9	8.2	4.1	1.2
Total	100.0	12.5	11.0	32.5	100.0	19.4	1.1	20.5	100.0	15.8	100.0

[1] C—Grade; [2] R—Recovery; [3] LREO (%)—light rare earth oxide (%); [4] HREO (%)—heavy rare earth oxide (%).

The rare-earth elements of each product were analyzed by ICP-MS, and the content of each element was expressed as an oxide form. These values were categorized and summarized as light rare-earth oxide (LREO) and heavy rare-earth oxide (HREO) on the basis of the periodic table. The TREO(%), the sum of LREO(%) and HREO(%), began to be measured at the magnetic intensity of 1.0 T or higher. This means that REE-bearing minerals were recovered between 1.0 T and 1.4 T. The TREO(%) by up to 48.0% at 1.2 T, and recovery of TREO(%) between 1.0 T and 1.4 T reached nearly 99%.

One of the characteristics of the analysis result is that the HREO(%) was significantly detected in the 1.0 T product. As monazite, the main REE-bearing mineral in feed sample, generally consists of light-REE, another REE-bearing mineral, a small amount of xenotime, was presumed to exist in this product as identified by MLA in Table 1. Therefore, mineral identification of the 1.0 T product was further conducted by XRD analysis, and the presence of xenotime was confirmed, as shown in Figure 4. The XRD peak of xenotime, which was not detected in Figure 2 due to the low content, was observed in the 1.0 T product, and ilmenite and monazite were observed as expected from the element analysis data of Table 4. Xenotime is typically found with monazite in concentrations of 0.5%–5.0%, and it is known to have strong magnetic characteristics as compared to monazite [17]. Therefore, xenotime was firstly recovered at 1.0 T, after which monazite was recovered at an intensity of above 1.2 T.

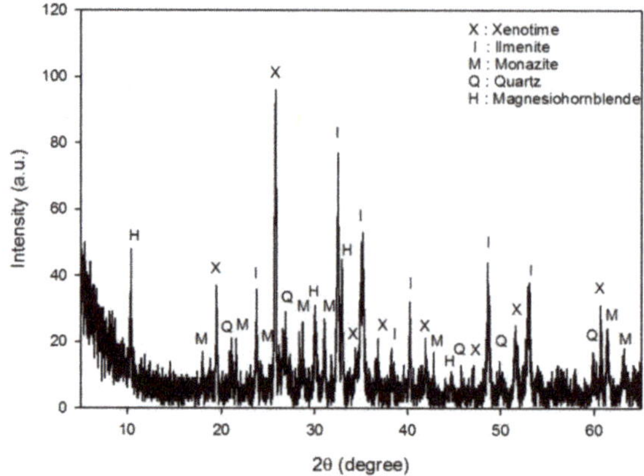

Figure 4. X-ray diffraction (XRD) pattern of 1.0 T magnetic product.

The analysis result of SiO_2 which is a constituent element of quartz showed that more than 75% of SiO_2 was distributed as a non-magnetic product. The SiO_2 content measured in the magnetic product was considered to be the magnetic minerals which are not fully liberated from the quartz, or the magnetic minerals containing the Si such as magnesio-hornblende as shown in Figure 4.

Figure 5 shows the cumulative recovery of ilmenite and TREO(%) as a function of magnetic intensity. At magnetic intensities of 0.6 T and 0.8 T, mainly ilmenite was recovered, and the cumulative recovery reached nearly 95% up to 0.8 T. On the other hand, at magnetic intensity above 1.2 T, the TREO(%) increased sharply and the cumulative recovery was nearly 99% between 1.0 T and 1.4 T. Therefore, based on magnetic intensities of 0.8 T and 1.4 T, ilmenite, monazite/xenotime, and non-magnetic minerals could be clearly separated via the magnetic separation process.

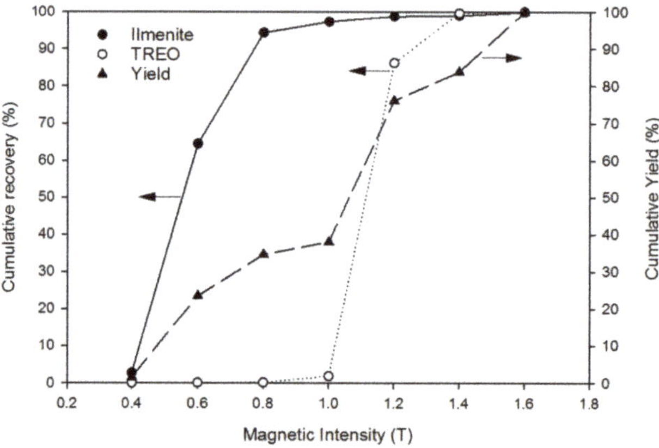

Figure 5. Cumulative recovery of ilmenite and TREO (%) as a function of magnetic intensity.

3.3. Batch Test for Beneficiation Process

Using a previous separation process result, 10 kg of batch test was conducted for the rare earth deposit as shown in Figure 6. The yield, grade (mineral content), and recovery of target elements including SiO_2, which is representative elements of gangue minerals, are shown in Table 5. According to the magnetic response characteristics, ilmenite was first recovered as a magnetic product below 0.8 T. Approximately 35.2% of the sample was recovered as ilmenite concentrate, and its calculated ilmenite grade and recovery were 90.9% and 98.5%, respectively. After the first magnetic separation process, the second magnetic separation process was conducted to recover REE-bearing xenotime and monazite. When the magnetic intensities ranged from 0.8 T and 1.4 T, approximately 44.5% of the sample was recovered, and the TREO(%) of this product was about 45.0%.

Table 5. The grade and recovery of main minerals at various magnetic intensity products.

Product		Yield (wt %)	Ilmenite				REE		Zircon		SiO$_2$	
			Fe (%)	Ti (%)	^{1} C (%)	^{2} R (%)	TREO (%)	^{2} R (%)	Zr (%)	^{2} R (%)	^{1} C (%)	^{2} R (%)
REEs	Conc.I	44.5	1.4	0.3	0.9	1.3	45.0	97.6	0.1	2.8	4.4	12.4
	Conc.II	0.9	0.4	0.2	0.6	<0.05	45.0	2.0	0.1	0.1	4.3	0.3
	Sub-total	45.4	1.4	0.3	0.9	1.3	45.0	99.6	0.1	2.9	4.4	12.7
Ilmenite		35.2	33.5	30.5	90.9	98.5	0.2	0.3	0.1	1.6	5.0	11.2
Zircon		2.1	1.5	4.3	0.9	<0.05	0.8	0.1	42.8	77.2	30.7	4.0
Gangue mineral		17.3	0.1	0.2	0.3	0.2	0.1	0.1	1.2	18.3	66.0	72.1
Total		100.0	12.5	11.0	32.5	100	20.5	100	1.1	100	15.8	100

^{1} C—Grade; ^{2} R—Recovery (%)

In general, zircon is included as a useful mineral along with rare-earth minerals in placer deposits. Therefore, in 10 kg of batch test, additional experiments were conducted to recover the zircon concentrate. Since zircon is a non-magnetic mineral, it was considered to be contained in 1.4 T non-magnetic product, and the gravity separation method was applied to separate zircon of high density (density = 4.6–4.7 g/cm^3) from low density gangue mineral such as quartz and orthoclase. Among various operating conditions of the shaking table, the optimum conditions were determined as follows: The angle of the shaking table was 2.5°, and shaking amplitude was 15 mm, and water flowrate was 10 L/min. However, the frequency was fixed at 300 rpm owing to the fixed motor speed. The shaking table separation removed 14.3% of the sample as light gangue minerals, which mainly consisted of SiO$_2$ and Al$_2$O$_3$. The yield of light gangue minerals from the shaking table was similar to that from float fraction in the sink-float test (Table 3), indicating that most light minerals were separated through the gravity separation method.

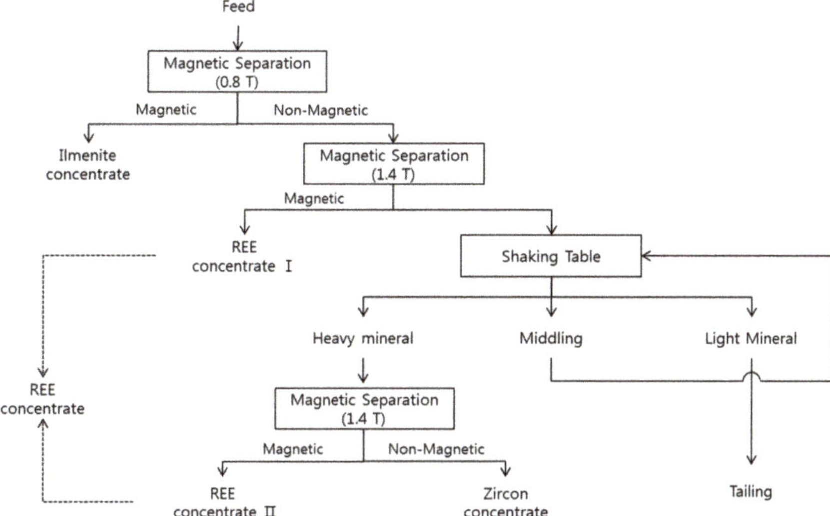

Figure 6. Flowchart of a batch test for rare earth element (REE) bearing placer deposits.

The high density product recovered from the non-magnetic product was conducted in the third magnetic separation process to recover the remaining rare-earth minerals. As a result, zircon could be recovered as a non-magnetic product in the third magnetic separation process, at a 2.1% yield, 42.8% grade, and 77.2% recovery. Additional REE-bearing minerals, with a TREO(%) of 45.0%, were recovered as magnetic products in the third magnetic separation process. Therefore, the total yield of REE concentrate was 45.4%, with a TREO of 45.0% and recovery of 99.6%.

Figure 7 shows the XRD pattern of the batch test results of each product. The detected minerals of each product agreed with the chemical composition analysis results in Table 5. In particular, the existence of minor minerals was not detected in ilmenite and REE concentrates, and it could be confirmed that the separation of paramagnetic minerals, ilmenite and monazite/xenotime, through the adjustment of magnetic intensity was good. In case of zircon concentrate, some anatase with high density (density = 3.8–4.0 g/cm^3) and quartz which was not fully separated were identified along with recovered zircon concentrates.

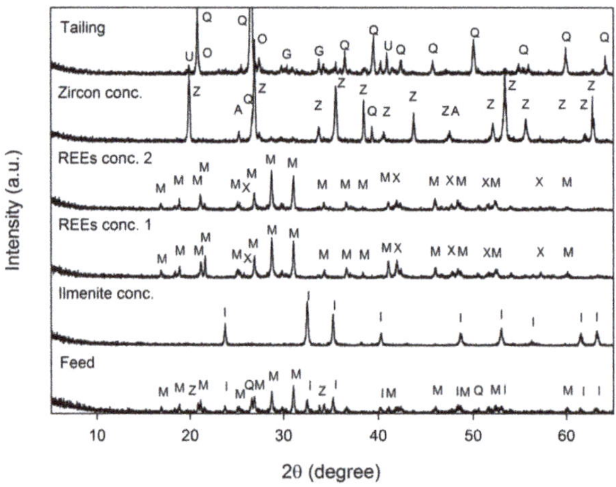

Figure 7. X-ray diffraction patterns of beneficiation process products. A—Anatase; G—Grossular; I—Ilmenite; M—Monazite; O—Orthoclase; Q—Quartz; U—Muscovite; X—Xenotime; Z—Zircon.

4. Conclusions

This paper describes the mineralogical characteristics of an REE-bearing placer deposit from North Korea and the steps taken to apply a beneficiation process using magnetic and gravity separation. The conclusions are as follows:

1. The feed sample was obtained from the bottom of the active river channel and sand bar of the Sam-Cheon area in North Korea. The Sam-Cheon area is located on the Rimjingang belt in the middle Korean Peninsula, and there are two REE mines, Wolbong mine and Ryeonsan mine, in the vicinity. The sample was preliminary concentrated in the mining site by gravity separation, and mineralogical analysis and beneficiation tests were conducted using preliminary concentrated heavy minerals sample.

2. The mineralogical analyses by XRD, XRF, ICP, and MLA indicated that the REE-bearing minerals in this sample were mostly monazite, and to a lesser extent, xenotime. Besides these REE-bearing minerals, ilmenite and zircon were valuable minerals to be concentrated.

3. Because this sample came from a placer deposit, the degree of liberation was sufficiently high. The "liberated" minerals, which have a mineral composition (wt %) of more than 80%, exceeded 90% for main minerals (monazite, ilmenite, and quartz).

4. Valuable minerals could be recovered via magnetic separation through various magnetic intensities. Ilmenite was recovered first between the magnetic intensities of 0.6 T and 0.8 T, and the REE-bearing minerals, xenotime and monazite were then recovered between 1.0 T and 1.4 T.

5. The 10 kg batch test was conducted to confirm the feasibility of the process using the unit separation result. As a result, the grade of ilmenite increased from 32.5 to 90.9%, and TREO(%) was enhanced from 20.5% to 45.0%. Additionally, zircon, another useful mineral, could be concentrated to 42.8% of the grade in heavy minerals of non-magnetic products. Consequently, it is confirmed that it was possible to separate valuable minerals selectively by simple magnetic separation and additional gravity separation.

Author Contributions: K.K. analyzed the data and wrote the paper and S.J. designed and performed the experiment.

Funding: This research was supported by the Convergence Research Project (CRC-15-06-KIGAM) funded by the National Research Council of Science and Technology (NST).

Conflicts of Interest: The authors declare no conflict of interest.

References

1. Gupta, C.; Krishnamurthy, N. Extractive metallurgy of rare earths. *Int. Mater. Rev.* **1992**, *37*, 197–248. [CrossRef]

2. Gupta, C.; Krishnamurthy, N. *Extractive Metallurgy of Rare Earths*; CRC Press: Boca Raton, FL, USA, 2005.

3. U.S. Geological Survey. Available online: https://pubs.er.usgs.gov/publication/70170140 (accessed on 25 February 2019).

4. Jordens, A.; Cheng, Y.P.; Waters, K.E. A review of the beneficiation of rare earth element bearing minerals. *Miner. Eng.* **2013**, *41*, 97–114. [CrossRef]

5. Cui, H.; Anderson, C.G. Alternative flowsheet for rare earth beneficiation of Bear Lodge ore. *Miner. Eng.* **2017**, *110*, 166–178. [CrossRef]

6. Jordens, A.; Sheridan, R.S.; Rowson, N.A.; Waters, K.E. Processing a rare earth mineral deposit using gravity and magnetic separation. *Miner. Eng.* **2014**, *62*, 9–18. [CrossRef]

7. Jordens, A.; Marion, C.; Langlois, R.; Grammatikopoulos, T.; Rowson, N.A.; Waters, K.E. Beneficiation of the Nechalacho rare earth deposit. Part 1: Gravity and magnetic separation. *Miner. Eng.* **2016**, *99*, 111–122. [CrossRef]

8. Jordens, A.; Marion, C.; Langlois, R.; Grammatikopoulos, T.; Sheridan, R.S.; Teng, C.; Demers, H.; Gauvin, R.; Rowson, N.A.; Waters, K.E. Beneficiation of the Nechalacho rare earth deposit. Part 2: Characterisation of products from gravity and magnetic separation. *Miner. Eng.* **2016**, *99*, 96–110. [CrossRef]

9. Li, L.Z.; Yang, X. China's rare earth ore deposits and beneficiation techniques. In Proceedings of the European Rare Earth Resource Conference, Milos Island, Greece, 4–7 September 2014.

10. Panda, R.; Kumari, A.; Jha, M.K.; Hait, J.; Kumar, V.; Kumar, J.R.; Lee, J.Y. Leaching of rare earth metals (REMs) from Korean monazite concentrate. *J. Ind. Eng. Chem.* **2014**, *20*, 2035–2042. [CrossRef]

11. Yang, X.; Satur, J.V.; Sanematsu, K.; Laukkanen, J.; Saastamoinen, T. Beneficiation studies of a complex REE ore. *Miner. Eng.* **2015**, *71*, 55–64. [CrossRef]

12. Moustafa, M.I.; Abdelfattah, N.A. Physical and chemical beneficiation of the Egyptian beach monazite. *Resour. Geol.* **2010**, *60*, 288–299. [CrossRef]

13. Rajan, G.R.; Sundararajan, M. Combined magnetic, electrostatic, and gravity separation techniques for recovering strategic heavy minerals from beach sands. *Mar. Georesour. Geotechnol.* **2018**, *36*, 959–965. [CrossRef]

14. Xiong, W.L.; Deng, J.; Chen, B.Y.; Deng, S.Z.; Wei, D.Z. Flotation-magnetic separation for the beneficiation of rare earth ores. *Miner. Eng.* **2018**, *119*, 49–56. [CrossRef]

15. Kang, M.S.; Choi, B.S.; Ryu, D.M.; Huh, D.H.; Lim, D.H.; Seo, E.R. *Chosun Geology Series 7*; Gong-Up Press: Pyongyang, Korea, 2011.

16. Kang, P.C.; Chwae, U.C.; Kim, K.B.; Hong, S.H.; Lee, B.J.; Hwang, J.H.; Park, K.H.; Hwang, S.K.; Choi, P.Y.; Song, K.Y. *Geological Map of Korea, Scale 1:1,000,000*; Korea Institute of Geology, Mining and Materials: Daejeon, Korea, 1995.
17. Kumari, A.; Panda, R.; Jha, M.K.; Kumar, J.R.; Lee, J.Y. Process development to recover rare earth metals from monazite mineral: A review. *Miner. Eng.* **2015**, *79*, 102–115. [CrossRef]

Article

Numerical Simulation of Flow Field Characteristics and Separation Performance Test of Multi-Product Hydrocyclone

Yuekan Zhang *, Peikun Liu, Lanyue Jiang, Xinghua Yang and Junru Yang

College of Mechanical & Electronic Engineering, Shandong University of Science and Technology, Qingdao 266590, China; lpk710128@163.com (P.L.); jianglanyue5@163.com (L.J.); yxh19781025@126.com (X.Y.); jryangzhang@163.com (J.Y.)
* Correspondence: zhangyk2007@sdust.edu.cn; Tel.: +86-0532-8605-7176

Received: 14 April 2019; Accepted: 15 May 2019; Published: 16 May 2019

Abstract: A traditional hydrocyclone can only generate two products with different size fractions after one classification, which does not meet the fine classification requirements for narrow size fractions. In order to achieve the fine classification, a multi-product hydrocyclone with double-overflow-pipe structure was designed in this study. In this work, numerical simulation and experimental test methods were used to study the internal flow field characteristics and distribution characteristics of the product size fraction. The simulation results showed that in contrast with the traditional single overflow pipe, there were two turns in the internal axial velocity direction of the hydrocyclone with the double-overflow-pipe structure. Meanwhile, the influence rule of the diameter of the underflow outlet on the flow field characteristics was obtained through numerical simulation. From the test, five products with different size fractions were obtained after one classification and the influence rule of the diameter of the underflow outlet on the size fraction distribution of multi-products was also obtained. This work provides a feasible research idea for obtaining the fine classification of multiple products.

Keywords: multi-product hydrocyclone; flow field characteristics; numerical simulation; experimental test

---

1. Introduction

A hydrocyclone is a representative device that utilizes the principle of centrifugal sedimentation to effectively separate two-phase or multi-phase liquid-liquid, liquid-solid, and liquid-gas mixtures having components of different densities [1,2]. It has many applications such as separation [3], sorting [4], liquid concentration [5], and liquid clarification [6]. The greatest advantage of the hydrocyclone is that unlike other centrifugal separation devices, no moving components are required. The separation process is completed by the fluid itself, which forms a vortex within the hydrocyclone. Hydrocyclones have the characteristics of high separation efficiency, low space requirements, large processing capacity, low separation cost, and continuous operation. Therefore, among the various solid-liquid separation technologies and equipment, the hydrocyclone is currently one of the most widely used equipment in industry. So far, the hydrocyclone is widely used in many industries such as mineral processing [7], petroleum [8], chemical industry [9], coal, mining [10], metallurgy [11], and tailings disposal [12].

In the rotating flow field of the hydrocyclone, under the condition of force balance, the bigger the particle diameter, the larger the radius of gyration. Thus, under the influence of the centrifugal force field, particles with different diameters follow a certain distribution rule along the radial direction inside the hydrocyclone. The coarse particles will be discharged from the underflow outlet with the external swirl, and the fine particles will be discharged from the overflow outlet with the internal swirl, thereby completing the classification of the coarse and fine particles.

From the perspective of practical application, the flow field study of the hydrocyclone does not seem important, because normally the focus is on the properties of the product obtained after hydrocyclone separation. As stated in the black-box theory, what matters in this case is usually the result, and not the process. However, in order to achieve greater separation efficiency and classification accuracy, it is necessary to improve the separation process, and the internal flow field of the hydrocyclone is an important factor that affects the separation process. Therefore, the importance of flow field research is self-evident. The flow field research, on the one hand, helps to understand the internal black-box theory inside the hydrocyclone and the separation mechanism of the hydrocyclone. On the other hand, the internal structure of the hydrocyclone can be improved and the influence rule of the structural parameters on the separation performance can be obtained.

The separation performance of the hydrocyclone can be improved by efficiently augmenting the structural parameters and form of the hydrocyclone [13]. Both experimental and theoretical studies by domestic and foreign scholars have generated a series of landmark research results in terms of research methods and content. In terms of the optimization of the structural parameters of the hydrocyclone, the representative studies include experimental studies on the hydrocyclone column height and diameter [14,15], feed inlet type and size [16,17], overflow pipe diameter, length, and shape, overflow pipe thickness [18–20], underflow outlet diameter, and shape, ratio relationship between the underflow outlet and overflow outlet [21–23]. A series of new hydrocyclone types have been designed, such as the built-in structural hydrocyclone [24,25], underflow outlet filled with flushing water [26], multi-stage series or parallel hydrocyclone [27–29], and three-product cyclone [30,31]. Mainza is one of the earliest researchers who put forward the three-product hydrocyclone, and the three-product hydrocyclone has been successfully tested in the Platinum industry for classifying UG2 ore which contains a high density chromite and a low density PGM carrying silica component. The emergence of these new technologies has promoted the further application of hydrocyclones in separation, however, all the aforementioned studies did not consider the influence of structural parameters on the separation efficiency.

Because the experiment is subject to different conditions, the numerical simulation method based on computational fluid dynamics (CFD) is getting more and more attention in the study of the internal flow field of the hydrocyclone. During the research, most scholars agree to use the Reynolds stress model (RSM) [32–36] to deal with the turbulence inside the hydrocyclone. For two-phase flow or multi-phase flow, most scholars prefer the discrete particle model (DPM) [37] to process the particle flow and use the volume of fluid (VOF) [38] model to process the gas-liquid contact surface, with the result that the simulation is consistent with the experiment is obtained.

Summarizing the latest domestic and foreign research progress on hydrocyclone structure, the consistent conclusion is that the structural parameters, especially the diameter of the underflow outlet, are the main factors that affect the separation performance of the flow field. However, most of the previous research focused on the separation performance, which could not overcome the shortcomings of existing hydrocyclones in which one classification cannot satisfy the requirements for fine classification with narrow size fraction. The conventional hydrocyclone can only obtain two products through the overflow of the fine particles and underflow of the coarse particles. However, in addition to these two products, there must be an intermediate product between the fine particles and the coarse particles. If the intermediate product enters the underflow, it will cause a loss in concentration. And if the intermediate product enters the overflow, it will cause the concentrate pollution. Therefore, effectively processing the intermediate material to obtain multiple products with narrow size fractions through a single classification that further meets the requirement for the fine classification of the feeding materials in the following sorting operation is key to improving concentrate yield and grade. Thus, a two-stage multi-product hydrocyclone that operates in series was designed in this work. The first stage of the hydrocyclone was designed as a coaxial double-overflow-pipe structure. The finest particle is discharged from the internal overflow pipe and the particle with the intermediate size is discharged from the external overflow pipe, which then relies on the residual pressure to enter the second stage of the hydrocyclone for subsequent fine grading. Thus, a single classification can obtain multiple products with different size fractions resulting from the first stage underflow, first stage overflow, second stage underflow, and second stage overflow. However, due to the special structure of the double-overflow-pipe, the vortex domain, boundary layer and flow regime change. Therefore, it is necessary to study the flow field performance. In this study, numerical analysis and experimental methods were used to study the internal flow field characteristics of the double-overflow-pipe hydrocyclone and the particle size distribution characteristics of the different products. The influence rule of the diameter of the underflow outlet on the flow field and particle size fraction distribution was also studied in this work.

2. Materials and Methods

2.1. Multi-Product Hydrocyclone

In this study, a two-stage series multi-product hydrocyclone shown in Figure 1 was designed. The first stage of the hydrocyclone is designed as a coaxial double-overflow-pipe structure with different diameters, that is, a smaller diameter overflow pipe is coaxially inserted into the overflow pipe of the conventional hydrocyclone. The second stage of the hydrocyclone is designed as a conventional structure with the upper part as the column section and the lower part as the cone section. The external overflow pipe of the first stage is connected to the feed inlet of the second stage through a pipeline.

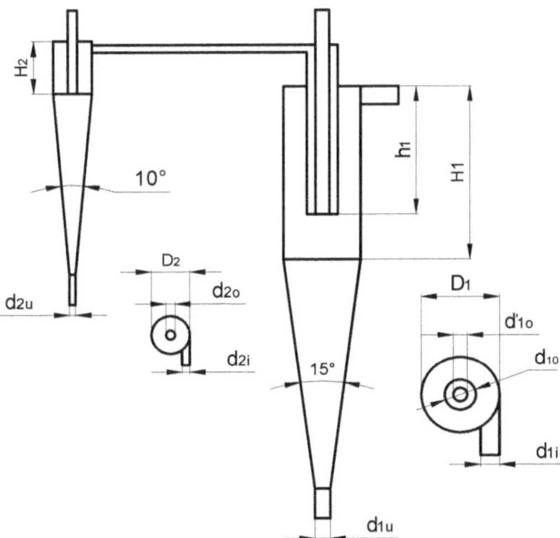

Figure 1. Schematic diagram of multi-product hydrocyclone.

During operation of the designed hydrocyclone, the slurry enters the first stage of the hydrocyclone at a certain tangential speed and the particle classification occurs under the influence of the centrifugal force. The coarsest particle is discharged from the underflow outlet, and the finest particle is discharged from the internal overflow pipe. The intermediate-sized particle is discharged from the external overflow pipe and then enters the second stage of the hydrocyclone where the fine classification continues under the influence of the residual pressure. Through this classification process, which involve the first stage underflow, first stage overflow, second stage underflow, and second stage overflow, multiple products with different size fractions are obtained. The structural dimensions of the hydrocyclone used in this paper are shown in Table 1.

Table 1. Specifications of the hydrocyclone used in this study.

Structural Parameter	Structural Dimensions (mm)
Hydrocyclone body diameter D_1	50
Inner vortex finder diameter d'_{1o}	6
Outer vortex finder diameter d_{1o}	20
Underflow port diameter d_{1u}	6, 8, 10, 12, 14
Feed inlet equivalent diameter d_{1i}	12
Outer overflow pipe insertion depth h_1	85
Hydrocyclone body diameter D_2	25
Vortex finder diameter d_{2o}	6
Underflow port diameter d_{2u}	4
Feed inlet equivalent diameter d_{2i}	5
Cylinder height H_1	116
Cylinder height H_2	35

2.2. Numerical Analysis Method

In this work, considering the double-overflow-pipe hydrocyclone as the research object, the fluid dynamics analysis software FLUENT 6.3 is used to simulate the flow field characteristics of the hydrocyclone to study the influence rule of the diameter of the underflow outlet on the velocity field and pressure field. We use the ICEM 14.5 software for structural meshing. The result of the meshing is shown in Figure 2 and the total number of nodes of the entire flow field in the calculation field is 140,577.

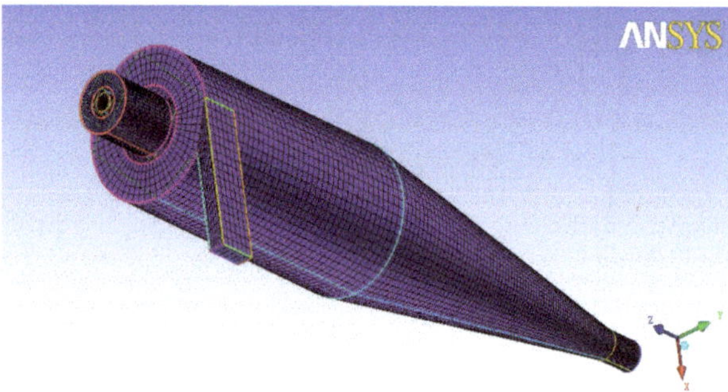

Figure 2. Mesh system used in the simulation.

The VOF two-phase flow model is used to represent the interface between air and water inside the hydrocyclone. The main phase is set as water and air is assumed to be the secondary phase. The Reynolds stress model (RSM) is used to represent the turbulence. A pressure-velocity coupling SIMPLE numerical method is used to calculate the control parameters. The pressure discretization format of the governing equation is the QUICK format. The fluid velocity at the inlet is 5 m/s, and it enters the hydrocyclone tangentially in a direction vertical to the inlet section. Set the overflow and underflow outlets as pressure outlets, and the wall of the hydrocyclone is represented by the standard wall function method. A pressure-based implicit transient 3D solver is used for the solution. The pressure gradient uses the Green Gaussian method to calculate the derivative term in the governing equation. The two-phase volume fraction uses the geo-reconstruct discrete format. The transient analysis uses explicit time discretization. The first order upwind scheme is adopted for modeling the turbulence kinetic energy, turbulence dissipation rate, and Reynolds stress discrete format.

2.3. Experiment Test Method

2.3.1. Experiment System

The experimental setup for testing the multi-product hydrocyclone is shown in Figure 3 and the schematic of the experiment is shown in Figure 4.

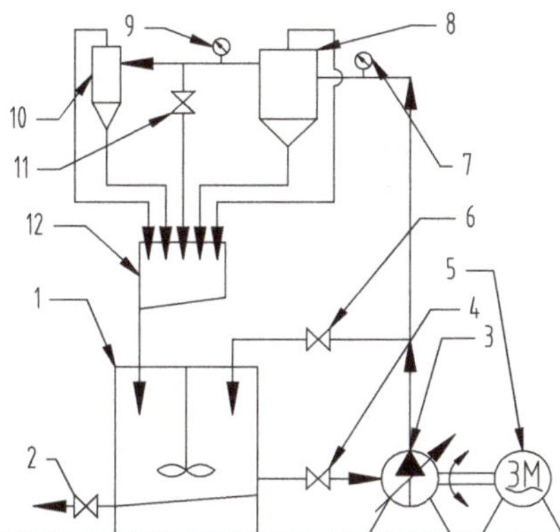

Figure 3. Schematic of the experimental apparatus. 1—Agitator; 2, 4, 6, 11—Valves; 3—Slurry pump; 5—Motor; 7, 9—Pressure gage; 8—I stage hycrocyclone; 10—II stage hydrocyclone; 12—Receiving vat.

Figure 4. Schematic of the experimental site.

It is mainly composed of a slurry tank, stirrer, frequency conversion slurry pump, flow metering unit, and pressure measuring unit. Rotor flowmeters are installed at the feed inlet and overflow outlet of the system to obtain the feed and overflow discharges. The underflow discharge is calculated indirectly from the feed discharge and the overflow discharge. Pointer-type precision pressure gauges are installed at the feed inlet, underflow outlet, and overflow outlet to measure the pressure values at each portion of the hydrocyclone. The flow and pressure in the experiment system are controlled and regulated by the valves installed in the pump and pipe. Sampling ports are arranged at the inlet and outlet of the hydrocyclone, and the material property analysis can be done at any instance during the experiment process.

2.3.2. Experiment Material

The material used in the experiment is fly ash, which is tested and analyzed by the laser particle size analyzer (Malvern Mastersizer 2000, Malvern, Worcestershire, UK). The particle size component is shown in Table 2. The proportion of the particles with diameters less than 10 μm (1250 mesh) is 44.13%.

The proportion of the particles with diameter less than 15 μm (800 mesh) is 53.11%. The proportion of the particles with diameter less than 44 μm (325 mesh) is 81.23%.

Table 2. Particle size composition of coal ash.

Mesh Number	Particle Size (μm)	Weight (%)	Cumulative Weight (%)
−70 + 100	−251 + 147	0.15	100
−100 + 200	−147 + 74	7.24	99.85
−200 + 270	−74 + 53	7.44	92.61
−270 + 325	−53 + 44	3.94	85.17
−325 + 450	−44 + 32	8.56	81.23
−450 + 800	−32 + 15	19.56	72.67
−800 + 1250	−15 + 10	8.98	53.11
−1250 + 2500	−10 + 5	15.03	44.13
−2500 + 6250	−5 + 2	14	29.1
−6250	−2 + 1	15.1	15.1

2.3.3. Experiment Design

The mass concentration of the fly ash slurry is taken to be 15%. During the experiment, the pressure gauge is adjusted through valves installed in the pump and pipe so that the feed pressure of the hydrocyclone is 0.16 MPa. The diameter of the internal overflow pipe of the hydrocyclone is 9 mm. The inserted depth of the internal overflow pipe is 85 mm. The diameter of the underflow outlet can be varied as 6 mm, 8 mm, 10 mm, 12 mm, and 14 mm. At the end of the experiment, the first stage internal overflow, first stage external overflow, first stage underflow, second stage overflow, and second stage underflow are sampled for the component analysis of the particle size fraction.

3. Results and Discussions

3.1. Numerical Simulation Results Analysis

3.1.1. Distribution Characteristics of Velocity Field and Influence of Underflow Outlet Diameter on Flow Field Performance

Figure 5 shows the tangential velocity distribution inside the hydrocyclone corresponding to different diameters of the underflow outlet. As shown in Figure 5, we can see that the tangential velocity at the center of the hydrocyclone is very small, and it increases with the radius of the hydrocyclone. When the diameter of the underflow outlet is reduced, the tangential velocity increases. This is because the smaller the diameter of the underflow outlet, the greater the resistance to the downward flow of the fluid and the lower the axial velocity. Therefore, when the feed pressure is kept constant, the tangential velocity is increased. From Figure 5, we can also conclude that the smaller the underflow outlet diameter, the closer the location of the maximum tangential velocity is to the center of the hydrocyclone; therefore, the greater the internal swirl centrifugal force, the better the separation performance and higher the classification accuracy.

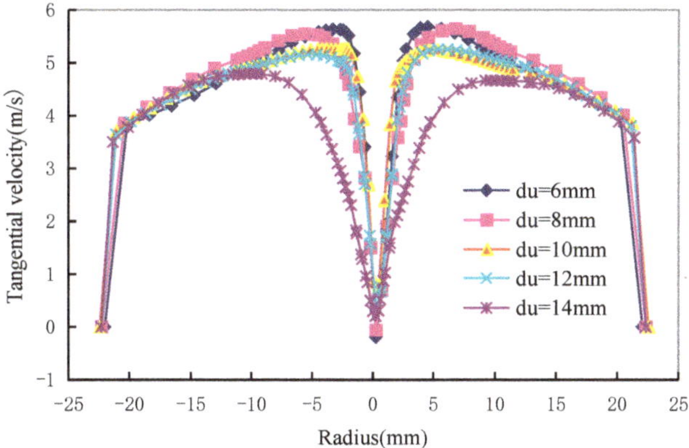

Figure 5. Tangential velocity profiles at several spigot diameters.

Figure 6 shows the axial velocity field distribution. It can be seen that the axial velocity is in the shape of a broken line wave. Near the wall of the hydrocyclone, the axial velocity is downward. As the radius of the hydrocyclone decreases, the axial velocity shows an upward trend and traverses the zero point. After reaching a maximum value, the axial velocity begins to show a downward trend and again passes the zero point. Thus, along the radial direction, the axial velocity passes through the zero point twice, that is, the direction changes twice. This is due to the special double-overflow-pipe structure. Because there are two coaxial overflow pipes, two upward internal swirls exist in the internal and external overflow pipes, causing the direction of the axial velocity to change twice. This is different from the traditional hydrocyclone with a single overflow pipe where the distribution of the axial velocity is the inverted "W".

As can be seen from Figure 6, in the double-overflow-pipe hydrocyclone, an area is formed between two zero axial velocity points. In this area, the axial velocity is small, which indicates that the position is not conducive to the separation of materials. Therefore, when designing a double-overflow-pipe hydrocyclone, we should try to avoid or reduce the range of this area by changing the structural parameters which will reduce the influence of this area on the separation efficiency of the hydrocyclone. From the influence rule of the diameter of the underflow outlet on the axial velocity, it can be seen that the axial velocity increases with the diameter of the underflow outlet. This indicates that increasing the diameter of the underflow outlet appropriately is beneficial to improving the classification efficiency.

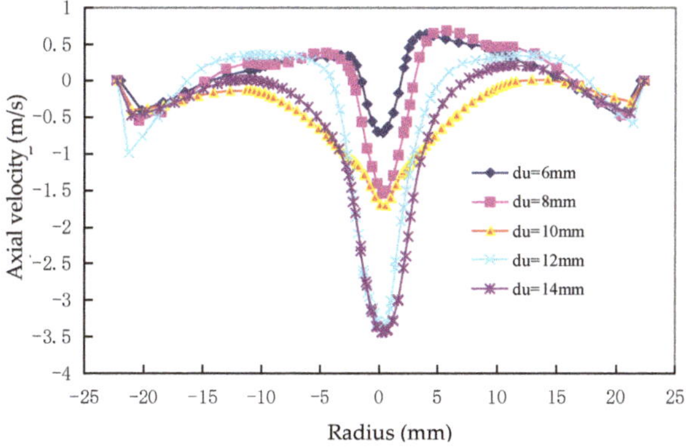

Figure 6. Axial velocity profiles at several underflow pipe diameters.

Figure 7 shows the radial velocity distribution inside the hydrocyclone. We can see that the radial velocity of the fluid is the smallest at the wall which is close to zero. As the radius decreases, the absolute value of the radial velocity gradually increases. After reaching the maximum value, it gradually shrinks with the decrease in the radius. This is, in essence, consistent with the conclusion of the research by Ji et al. [39]. The diameter of the underflow outlet has little effect on the radial velocity field, which demonstrates that there is no essential difference between the double-overflow-pipe structure and the traditional single-overflow-pipe structure in terms of the radial velocity distribution.

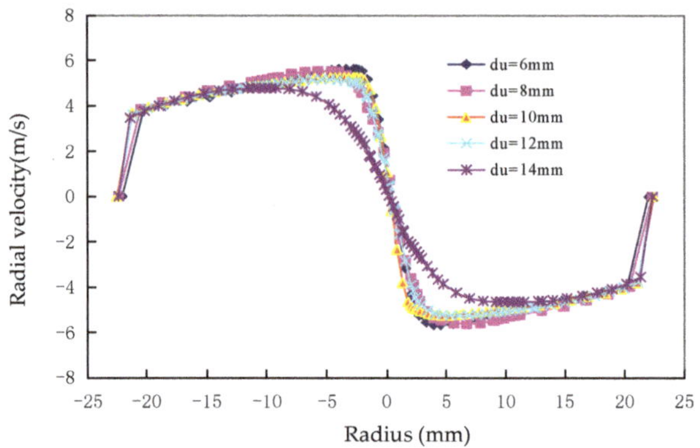

Figure 7. Radial velocity profiles at several underflow pipe diameters.

3.1.2. Distribution Characteristics of Pressure Field and Influence of Underflow Outlet Diameter on Pressure Field Performance

Figure 8 shows the internal pressure field distribution of the hydrocyclone with double-overflow-pipe structure. The internal pressure decreases gradually from the wall to the axis, and it is symmetrically distributed at the center. The pressure near the axial direction is zero, and the nearer the point considered is to the central axis, the greater the negative pressure. From the simulation results, we can see that the hydrocyclone with double-overflow-pipe structure is very similar to the traditional hydrocyclone with a single-overflow-pipe structure in terms of the pressure distribution. We can also observe that the diameter of the underflow outlet has little effect on the pressure field. As the diameter of the underflow outlet increases, the pressure decreases slightly, but the change is not obvious.

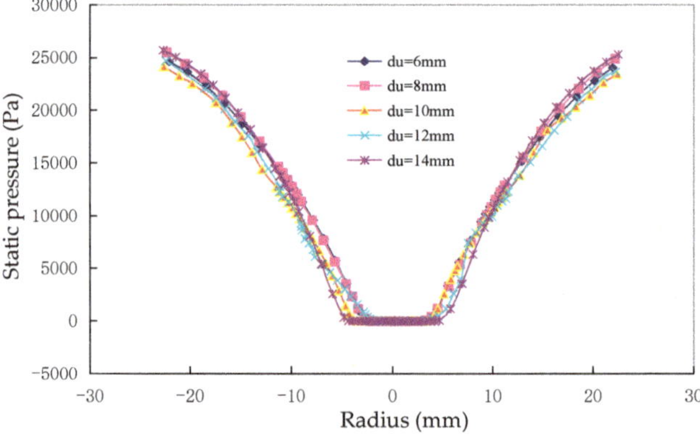

Figure 8. Static pressure profiles at several underflow pipe diameters.

3.2. Experiment Results Analysis

3.2.1. Influence of Underflow Outlet Diameter on Particle Size Fraction Component of First Stage Internal Overflow

Table 3 shows the particle size fraction component of the first stage internal overflow for different diameters of the underflow outlet, which are 6 mm, 8 mm, 10 mm, 12 mm, and 14 mm. From the Table 3, we can see that the internal overflow particle size decreases when the diameter of the underflow outlet decreases. When the diameter of the underflow outlet is 14 mm, $D98 = 44.13$ μm. When the diameter is 6 mm, $D98 = 36.91$ μm. The results show that proper reduction of the diameter of the underflow outlet is beneficial to obtain the first stage internal overflow product with finer particle size fraction.

Table 3. Particle size of I inner overflow.

Underflow Diameter (mm)	6	8	10	12	14	Content (%)
	0.512	0.521	0.532	0.53	0.562	3
	0.821	0.865	0.86	0.888	0.974	10
Particle size (μm)	3.523	3.896	4.105	4.266	5.378	50
	16.35	16.98	18.65	18.65	23.83	90
	36.91	37.14	37.9	38.14	44.13	98

3.2.2. Influence of Underflow Outlet Diameter on the Particle Size Fraction Component of First Stage External Overflow

Table 4 shows the particle size fraction component of the first stage external overflow with different diameters of the underflow outlet, which are 6 mm, 8 mm, 10 mm, 12 mm, and 14 mm. We can see from the Table 4 that the particle size of the first stage external overflow has a tendency to become thinner when the diameter of the underflow outlet increases. When the diameter of the underflow outlet is 6 mm, $D98 = 51.23$ μm. When the diameter is 14mm, $D98 = 45.05$ μm. This indicates that, for a multi-product hydrocyclone with the same specification, a proper reduction in the diameter of the underflow outlet is beneficial to obtain the first stage external overflow product with coarser particle size fraction.

Table 4. Particle size of I outer overflow.

Underflow Diameter (mm)	6	8	10	12	14	Content (%)
	0.554	0.554	0.557	0.557	0.547	3
	0.974	0.967	0.95	0.97	0.935	10
Particle size (μm)	5.842	5.476	5.209	5.083	4.969	50
	27.2	24.46	22.95	22.35	22.25	90
	51.23	47.64	46.05	45.45	45.05	98

Comparing Table 4 with Table 3, we can conclude that for the same diameter of the underflow outlet, the first stage external overflow particle size is slightly coarser than the first stage internal overflow particle size. This indicates that for a double-overflow-pipe hydrocyclone, we can obtain internal and external overflow products with two different particle size fractions. Meanwhile, we can see that the diameter of the underflow outlet has an opposite effect on the internal and external overflow particle sizes. That is, as the diameter of the underflow outlet increases, the first stage internal overflow particle size becomes coarser and the first stage external overflow particle size becomes finer. Based on this, we can adjust the diameter of the underflow outlet to obtain the internal and external overflow products having different particle sizes, which can further meet the requirements of the subsequent sorting operations on the particle size of the feeding materials.

3.2.3. The Influence of the Diameter of the Underflow Outlet on the Particle Size Fraction Component of the First Stage Underflow

Table 5 shows the particle size fraction component of the first stage underflow with different diameters of the underflow outlet, which are 6 mm, 8 mm, 10 mm, 12 mm, and 14 mm. From Table 5, we can see that the first stage underflow particle becomes finer when the diameter of the underflow outlet increases. When the diameter of the underflow outlet is 6 mm, D98 = 137.1 μm. When the diameter is 14 mm, D98 = 127 μm. This indicates that the proper reduction of the diameter of the underflow outlet is beneficial to obtain the first stage underflow product with coarser particle size fractions.

Table 5. Particle size of I underflow.

Underflow Diameter (mm)	6	8	10	12	14	Content (%)
	0.854	0.86	0.748	0.723	0.679	3
	3.334	2.807	2.206	2.02	1.814	10
Particle size (μm)	32.22	29.61	27	24.34	21.88	50
	94.8	90.56	90.15	86.01	83.36	90
	137.1	132.4	132.8	128.6	127	98

3.2.4. Influence of the Underflow Outlet Diameter on the Particle Size Fraction Component of Second Stage Overflow

Table 6 shows the particle size fraction component of the second stage overflow with different diameters of the underflow outlet which are 6 mm, 8 mm, 10 mm, 12 mm, and 14 mm. It can be seen from Table 6 that the change in the diameter of the underflow outlet has little effect on the particle size of the second stage overflow. When the diameter of the underflow outlet is increased from 6 mm to 8 mm, the particle size gradually becomes fine. On the other hand, when the diameter of the underflow outlet is increased from 10 mm to 14 mm, as the diameter increases, the particle gradually becomes coarse. This is because the first stage external overflow, which is the second stage feed material, becomes finer with the increase in the underflow outlet diameter. Generally speaking, the second stage overflow should show the same trend. However, in the practical experiment test, the second stage pressure decreases with the increase of the diameter of the underflow outlet, and the decrease of the pressure makes the second stage overflow become coarser. So, we can conclude that it is the result of the combination of these two factors.

Table 6. Particle size of II overflow.

Underflow Diameter (mm)	6	8	10	12	14	Content (%)
	0.55	0.553	0.534	0.544	0.546	3
	0.975	0.932	0.89	0.89	0.928	10
Particle size (μm)	5.352	5.016	4.476	4.671	4.677	50
	25.57	22.31	20.29	21.02	22.01	90
	46.3	45.26	43.81	43.98	44.95	98

3.2.5. Influence of Underflow Outlet Diameter on the Particle Size Fraction Component of Second Stage Underflow

Table 7 shows the particle size fraction component of the second stage underflow with different diameters of the underflow outlet which are 6 mm, 8 mm, 10 mm, 12 mm, and 14 mm. From Table 7 we can see that the second stage underflow particle size becomes finer when the diameter of the underflow outlet increases. When the diameter of the underflow outlet is 6 mm, D98 = 63.32 μm. When the diameter is 14 mm, D98 = 51.72 μm.

Table 7. Particle size of II underflow.

Underflow Diameter (mm)	6	8	10	12	14	Content (%)
	0.558	0.555	0.549	0.539	0.54	3
	1.198	1.05	1.1	1.025	1.017	10
Particle size (μm)	8.213	7.684	6.721	6.709	6.66	50
	32.76	29.55	29.2	28.94	27.61	90
	63.32	58.26	55.16	53.69	51.72	98

The size distribution results have shown that five distinct products can be produced by the two-stage series multi-product hydrocyclone, which is consistent with Mainza's results [30]. The difference is that we think that the diameter of first underflow port is an important factor affecting the size distribution, while Mainza et al. believe that the depth of insertion of overflow pipe is an important factor affecting the size distribution. The reason is that the pressure of the second stage of the two-stage multi-product hydrocyclone comes from the residual pressure of the first stage, and the diameter of the underflow port has a greater impact on the pressure.

3.2.6. Comparison of Particle Size Fraction Component of Different Products

Figure 9 shows the comparison of the maximum particle diameters of four products obtained from the first stage overflow, first stage underflow, second stage overflow, and second stage underflow. We can see that the proposed two-stage series multi-product hydrocyclone can obtain the products with different particle diameters after the separation. The maximum particle diameter of the first stage internal overflow is 44.13 μm. The maximum particle diameter of the first stage underflow is 127 μm. The maximum particle diameter of the second stage overflow is 44.95 μm. The maximum particle diameter of the second stage underflow is 51.72 μm.

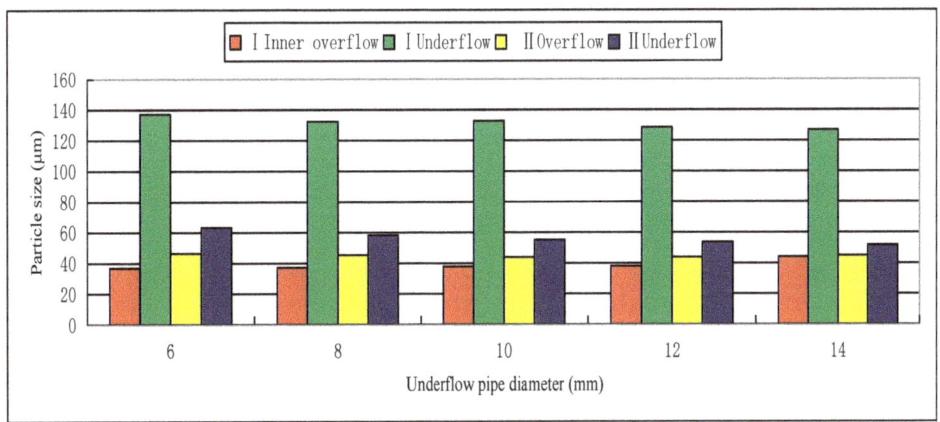

Figure 9. Comparison of maximum grain size products of different underflow pipe.

4. Conclusions

In this study, a two-stage series multi-product hydrocyclone was designed so that different particle size fractions could be obtained as the first stage internal overflow, first stage external overflow, first stage underflow, second stage overflow, and second stage underflow after one classification.

The axial velocity of the double-overflow-pipe changed direction twice along the radial direction, thus enclosing an area between the coaxial overflow pipes. The axial velocity was small in this area, which was not conducive to the particle separation. The diameter of the underflow outlet had little effect on the pressure field and radial velocity field, but had a greater influence on the tangential velocity and axial velocity. The axial velocity increased with the diameter of the underflow, which indicated that an optimal increase in the diameter of the underflow was beneficial to improving the classification efficiency. The tangential velocity increased with the decrease in the diameter of the underflow outlet, which was beneficial for increasing the centrifugal force of the internal swirl of the hydrocyclone, reducing the separation granularity, and improving the classification accuracy of the fine particles.

The first stage internal overflow was the finest while the first stage external overflow was the coarsest. The particle size of the second stage overflow was between that of the first stage internal and external overflows. The diameter of the underflow outlet had an opposite influence on the particle size of the first stage internal and external overflows. When the diameter of the underflow outlet increased, the particles of the first stage internal overflow became coarser and the particles of the first stage external overflow became finer.

The results of this study have a certain guiding role in the study of the flow field characteristics of the multi-product hydrocyclone. However, there are still many aspects that need to be addressed, such as the influence rule of the overflow pipe diameter and depth of insertion of the overflow pipe on the flow field and the distribution of the particle size fraction. In addition, the scale-up in engineering application is also an important issue to be discussed and highlight in the next step. These can be used to obtain the optimal structural parameters of the hydrocyclone for the best separation performance.

Author Contributions: Conceptualization, Y.Z.; Data curation, X.Y.; Formal analysis, Y.Z. and P.L.; Investigation, Y.Z. and L.J.; Methodology, Y.Z., P.L. and J.Y.; Resources, Y.Z.; Software, Y.Z.; Validation, L.J. and X.Y.; Writing—original draft, Y.Z.; Writing—review & editing, X.Y. and J.Y.

Funding: This research was funded by National Key R&D Program of China, 2018YFC0604702, Natural Science Foundation of Shandong province, ZR2016EEM37 and key research and development project of Shangdong province, 2017GSF216004.

Conflicts of Interest: The authors declare no conflict of interest.

References

1. Svarovsky, L. *Hydrocyclones*; Holt, Rinehart and Winston: New York, NY, USA, 1984.
2. Svarovsky, L. *Solid-liquid Separation*; Butterworths: London, UK, 1981.
3. Narasimha, M.; Sripriya, R.; Banerjee, P.K. CFD modeling of hydrocyclone-prediction of cut size. *Int. J. Miner. Process.* **2005**, *75*, 53–68. [CrossRef]
4. Liu, L.; Zhao, L.; Yang, X.; Wang, Y.; Xu, B.; Liang, B. Innovative design and study of an oil-water coupling separation magnetic hydrocyclone. *Sep. Purif. Technol.* **2019**, *213*, 389–400. [CrossRef]
5. Silva, D.O.; Vieira, L.G.M.; Barrozo, M.A.S. Optimization of design and performance of solid-liquid separators: A thickener hydrocyclone. *Chem. Eng. Technol.* **2015**, *38*, 319–326. [CrossRef]
6. Liu, P.-K.; Chu, L.-Y.; Wang, J.; Yu, Y.-F. Enhancement of Hydrocyclone Classification Efficiency for Fine Particles by Introducing a Volute Chamber with a Pre-Sedimentation Function. *Chem. Eng. Technol.* **2008**, *31*, 474–478. [CrossRef]
7. Narasimha, M.; Mainza, A.N.; Holtham, P.N.; Powell, M.S.; Brennan, M.S. A semi-mechanistic model of hydrocyclones Developed from industrial data and inputs from CFD. *Int. J. Miner. Process.* **2014**, *133*, 1–12. [CrossRef]
8. Huang, L.; Deng, S.; Guan, J.; Chen, M.; Hua, W. Development of a novel high-efficiency dynamic hydrocyclone for oil-water separation. *Chem. Eng. Res. Des.* **2018**, *130*, 266–273. [CrossRef]
9. Xu, Y.-X.; Liu, Y.; Zhang, Y.-H.; Yang, X.-J.; Wang, H.-L. Effect of shear stress on deoiling of oil-contaminated catalysts in a hydrocyclone. *Chem. Eng. Technol.* **2016**, *39*, 567–575. [CrossRef]
10. Chu, K.; Chen, J.; Yu, A.B.; Williams, R.A. Numerical studies of multiphase flow and separation performance of natural medium cyclones for recovering waste coal. *Powder Technol.* **2017**, *314*, 532–541. [CrossRef]
11. Vehmaanperä, P.; Safonov, D.; Kinnarinen, T.; Häkkinen, A. Improvement of the filtration characteristics of calcite slurry by hydrocyclone classifification. *Miner. Eng.* **2018**, *128*, 133–140. [CrossRef]
12. Mackay, I.; Mendez, E.; Molina, I.; Videla, A.R.; Cilliers, J.J.; Brito-Parada, P.R. Dynamic froth stability of copper flotation tailings. *Miner. Eng.* **2018**, *124*, 103–107. [CrossRef]

13. Li, Y.; Liu, C.; Zhang, T.; Li, D.; Zheng, L. Experimental and numerical study of a hydrocyclone with the modification of geometrical structure. *Can. J. Chem. Eng.* **2018**, *96*, 2638–2649. [CrossRef]

14. Neesse, T.; Dueck, J.; Schwemmer, H.; Farghaly, M. Using a high pressure hydrocyclone for solids classification in the submicron Range. *Miner. Eng.* **2015**, *71*, 85–88. [CrossRef]

15. Yang, Q.; Lv, W.-J.; Shi, L.; Wang, H.-L. Treating methanol-to-olefin quench water by mini hydrocyclone clarification and steam stripper purification. *Chem. Eng. Technol.* **2015**, *38*, 547–552. [CrossRef]

16. Vieira, L.G.M.; Silva, D.O.; Barrozo, M.A.S. Effect of inlet diameter on the performance of a filtering hydrocyclone separator. *Chem. Eng. Technol.* **2016**, *39*, 1406–1412. [CrossRef]

17. Zhang, C.; Wei, D.; Cui, B.; Li, T.; Luo, N. Effects of curvature radius on separation behaviors of the hydrocyclone with a tangent-circle inlet. *Powder Technol.* **2017**, *305*, 156–165. [CrossRef]

18. He, F.; Zhang, Y.; Wang, J.; Yang, Q.; Wang, H.; Tan, Y.H. Flow patterns in mini-hydrocyclones with different vortex finder depths. *Chem. Eng. Technol.* **2013**, *36*, 1935–1942. [CrossRef]

19. Ni, L.; Tian, J.; Zhao, J. Experimental study of the relationship between separation performance and lengths of vortex finder of a novel de-foulant hydrocyclone with continuous underflow and reflux function. *Sep. Sci. Technol.* **2016**, *52*, 142–154. [CrossRef]

20. Wang, B.; Yu, A.B. Numerical study of the gas-liquid-solid flow in hydrocyclones with different configuration of vortex finder. *Chem. Eng. J.* **2008**, *135*, 33–42. [CrossRef]

21. Ni, L.; Tian, J.; Zhao, J. Experimental study of the effect of underflow pipe diameter on separation performance of a novel defoulant hydrocyclone with continuous underflow and reflux function. *Sep. Purif. Technol.* **2016**, *171*, 270–279. [CrossRef]

22. Ghodrat, M.; Kuang, S.B.; Yu, A.B.; Vince, A.; Barnett, G.D.; Barnett, P.J. Computational study of the multiphase flow and performance of hydrocyclones: Effects of cyclone size and spigot diameter. *Ind. Eng. Chem. Res.* **2013**, *52*, 16019–16031. [CrossRef]

23. Kılavuz, F.Ş.; Gülsoy, Ö.Y. The effect of cone ratio on the separation efficiency of small diameter hydrocyclones. *Int. J. Miner. Process.* **2011**, *98*, 163–167. [CrossRef]

24. Zou, J.; Wang, C.; Ji, C. Experimental study on the air core in a hydrocyclone. *Dry. Technol.* **2016**, *34*, 854–860. [CrossRef]

25. Zhao, L.; Jiang, M.; Xu, B.; Zhu, B. Development of a new type high-efficient inner-cone hydrocyclone. *Chem. Eng. Res. Des.* **2012**, *90*, 2129–2134. [CrossRef]

26. Golyk, V.; Huber, S.; Farghaly, M.G.; Prolss, G.; Endres, E.; Neesse, T.; Hararah, M.A. Higher kaolin recovery with a water injection cyclone. *Miner. Eng.* **2011**, *24*, 98–101. [CrossRef]

27. Yang, G.; He, L.; Shujun, B.; Fengshan, W.; Guoxing, Z.; Iei, Z.; Bo, Y.; Li, Z. Analysis and test of flow field of two-stage series downhole hydrocyclone. In Proceedings of the SPE Asia Pacific Oil & Gas Conference and Exhibition, Perth, Australia, 25–27 October 2016. [CrossRef]

28. Huang, C.; Wang, J.-G.; Wang, J.-Y.; Chen, C.; Wang, H.-L. Pressure drop and flow distribution in a mini-hydrocyclone group: UU-type parallel arrangement. *Sep. Purif. Technol.* **2013**, *103*, 139–150. [CrossRef]

29. Chen, C.; Wang, H.-L.; Gan, G.-H.; Wang, J.-Y.; Huang, C. Pressure drop and flow distribution in a group of parallel hydrocyclones: Z-Z-type arrangement. *Sep. Purif. Technol.* **2013**, *108*, 15–27. [CrossRef]

30. Mainza, A.; Powell, M.S.; Knopjes, B. Differential classification of dense material in a three-product cyclone. *Miner. Eng.* **2004**, *17*, 573–579. [CrossRef]

31. Mainza, A.; Narasimha, M.; Powell, M.S.; Holtham, P.N.; Brennan, M. Study of flow behaviour in a three-product cyclone using computational fluid dynamics. *Miner. Eng.* **2006**, *19*, 1048–1058. [CrossRef]

32. Mokni, I.; Dhaouad, H.; Bournot, P.; Mhiri, H. Numerical investigation of the effect of the cylindrical height on separation performances of uniflow hydrocyclone. *Chem. Eng. Sci.* **2015**, *122*, 500–513. [CrossRef]

33. Vakamalla, T.R.; Koruprolu, V.B.R.; Arugonda, R.; Mangadoddy, N. Development of novel hydrocyclone designs for improved fines classification using multiphase CFD model. *Sep. Purif. Technol.* **2017**, *175*, 481–497. [CrossRef]

34. Huang, A.-N.; Ito, K.; Fukasawa, T.; Yoshida, H.; Kuo, H.-P.; Fukui, K. Classification performance analysis of a novel cyclone with a slit on the conical part by CFD simulation. *Sep. Purif. Technol.* **2018**, *190*, 25–32. [CrossRef]

35. Zhou, F.; Sun, G.; Zhang, Y.; Ci, H.; Wei, Q. Experimental and CFD study on the effects of surface roughness on cyclone performance. *Sep. Purif. Technol.* **2018**, *193*, 175–183. [CrossRef]

36. Misiulia, D.; Elsayed, K.; Andersson, A.G. Geometry optimization of a deswirler for cyclone separator in terms of pressure drop using CFD and artificial neural network. *Sep. Purif. Technol.* **2017**, *185*, 10–23. [CrossRef]

37. Murthy, Y.R.; Bhaskar, K.U. Parametric CFD studies on hydrocyclone. *Powder Technol.* **2012**, *230*, 36–47. [CrossRef]

38. Vakamalla, T.R.; Mangadoddy, N. Numerical simulation of industrial hydrocyclones performance: Role of turbulence modelling. *Sep. Purif. Technol.* **2017**, *176*, 23–39. [CrossRef]

39. Ji, L.; Kuang, S.; Qi, Z.; Wang, Y.; Chen, J.; Yu, A. Computational analysis and optimization of hydrocyclone size to mitigate adverse effect of particle density. *Sep. Purif. Technol.* **2017**, *174*, 251–263. [CrossRef]

Article

The Study on Numerical Simulation and Experiments of Four Product Hydrocyclone with Double Vortex Finders

Yuekan Zhang *, Peikun Liu, Lanyue Jiang and Xinghua Yang

College of Mechanical & Electronic Engineering, Shandong University of Science and Technology, Qingdao 266590, China; lpk710128@163.com (P.L.); jianglanyue5@163.com (L.J.); yxh2000@sina.com (X.Y.)

* Correspondence: zhangyk2007@sdust.edu.cn; Tel.: +86-0532-8605-7176

Received: 14 November 2018; Accepted: 24 December 2018; Published: 30 December 2018

Abstract: A hydrocyclone is an instrument that can effectively separate multi-phase mixtures of particles with different densities or sizes based on centrifugal sedimentation principles. However, conventional hydrocyclones lead to two products only, resulting in an over-wide particle size range that does not meet the requirements of subsequent operations. In this article, a two-stage series, a four product hydrocyclone is proposed. The first stage hydrocyclone is designed to be a coaxial double overflow pipe: under the effect of separation, fine particles are discharged from the internal overflow pipe, while medium-size particles are discharged from external overflow pipe before entering the second stage hydrocyclone for fine sedimentation. In other words, one-stage grading leads to four products, including the first stage underflow, the first stage overflow, the second stage underflow, and the second stage overflow. The effects of structural parameters and operational parameters on flow field distribution in hydrocyclone were investigated via a study of flow field distribution in multi-product hydrocyclones using numerical simulations. The application of four product hydrocyclone in iron recovery shows that the grade and recovery of iron concentrate exceed 65.08% and 86.14%, respectively. This study provides references for understanding the flow field distribution in hydrocyclones and development of multi-product grading instrument in terms of both theory and industrial applications.

Keywords: four product hydrocyclone with double vortex finders; separation; numerical simulation; experiments

---

1. Introduction

A hydrocyclone is an instrument that can effectively separate mixtures of particles with different density or size based on centrifugal sedimentation principles [1–7]. Once pressurized into a hydrocyclone, powders are separated by separation: coarse/dense particles shift to the sidewall due to relatively large centrifugal forces, join the underflow via an outer swirl and leave via an underflow outlet; fine/sparse particles shift to the core due to relatively small centrifugal forces, join the overflow via an inner swirl and leave via an overflow pipe. One-stage grading by conventional hydrocyclones leads to coarse particle underflow and fine particle overflow only [8–13]. As shown in Figure 1, the size range is over-wide and fine grading has not been achieved, resulting in poor grading efficiency and accuracy.

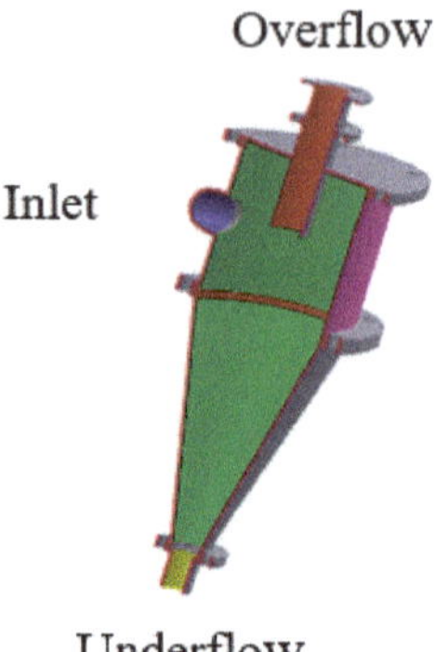

Figure 1. Traditional cyclone with two products.

To obtain multiple products in narrow size ranges using hydrocyclones, overflow series systems are usually employed. In particular, the first stage overflow serves as a feedstock for the second stage hydrocyclone, thus the fine particles mesh is obtained. By repeating this process, the multiple products in narrow size ranges may be produced, as shown in Figure 2.

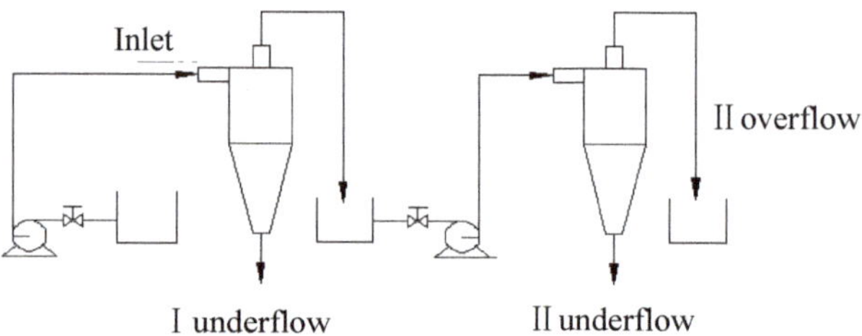

Figure 2. Classification diagram of the three product cycolone.

However, this technique is limited by a tedious process and costly equipment. Great efforts have been made to achieve multiple products in narrow size ranges by one-stage grading in hydrocyclones. For instance, a sieve hydrocyclone was proposed for grading of fine slurry particles (see Figure 3). Coarse particle underflow, fine particle overflow, and undersieve medium particle can be obtained. This approach, however, is limited by the mesh blocking issue.

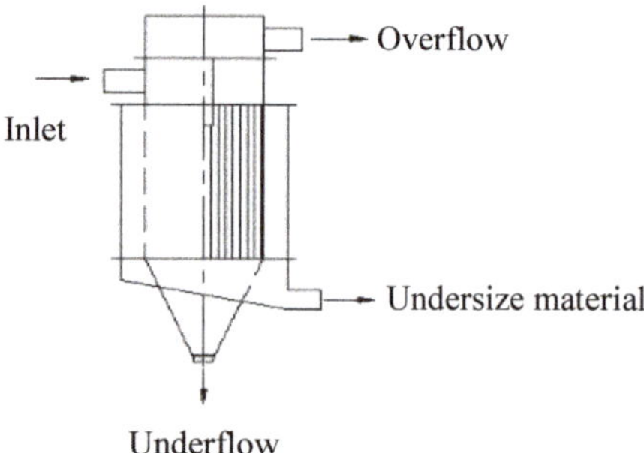

Figure 3. Diagram of the hydrocyclone with screen.

Ahmed et al. proposed a hydrocyclone for three products (one overflow and two underflows) by design of a tangential discharge outlet in a middle part of the cone section (see Figure 4) [14]. This three product hydrocyclone separates particles in sizes larger than the overflow fine particles, yet smaller than the underflow coarse particles. However, the size ranges of particles obtained by this process cannot be precisely controlled, thus so far it is not suitable for industrial applications.

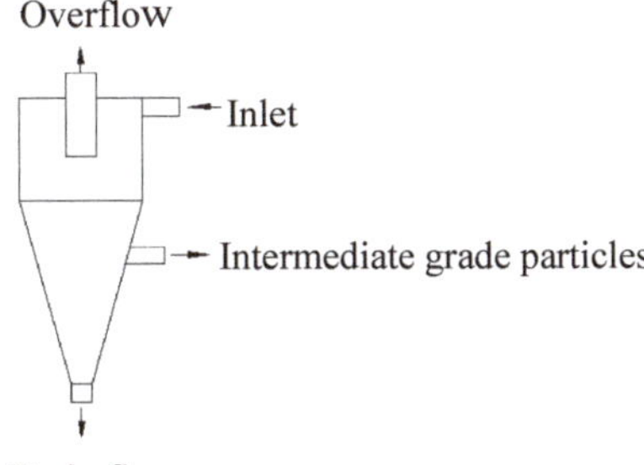

Figure 4. Three-product cyclone with double spigot.

A double overflow pipe three product hydrocyclone, with an insert of the other overflow pipe into the conventional hydrocyclone overflow pipe, was proposed in references [15–18]. In this way, the one-stage grading leads to an overflow containing fine particles, an overflow containing medium particles, and an underflow containing coarse particles. Nevertheless, the size ranges of particles in external overflow are still relatively wide and the diameter mismatch between internal and external overflow pipes can be a severe issue. Meanwhile, the effect of the insertion depth of an internal overflow pipe and effects of structural and functional parameters on the flow field performance have not been well understood.

In summary, the expanding applications of hydrocyclones lead to higher requirements on their grading accuracy and grading size. The one-stage grading by conventional hydrocyclones leads to two products only and the particle size ranges are over-wide. In such a case incomplete separation is inevitable, resulting in poor grading efficiency and accuracy and the phenomenon of "fine particle in underflow, coarse particle in overflow". In this study a two-stage series, a four product hydrocyclone (see Figure 5) is proposed.

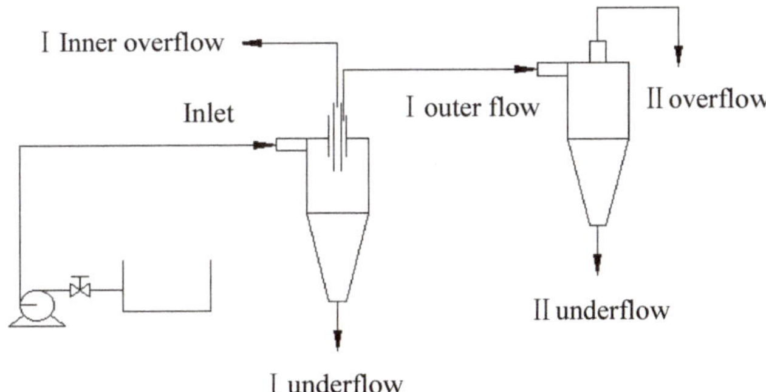

Figure 5. Four product cyclone with double vortex finders.

The first section hydrocyclone is designed as a coaxial double overflow pipe. Once pressurized into a hydrocyclone, matters are separated by separation: fine particles are discharged from the internal overflow pipe while medium-size particles are discharged from the external overflow pipe and serve as a feedstock to the second section hydrocyclone in which these particles are graded again. In this way four products, namely the first stage underflow, first stage overflow, the second stage underflow and the second stage overflow, can be obtained by one-stage grading. One-stage grading by the multi-product hydrocyclones leads to multiple products in narrow size ranges and is characterized by reduced particle size, improved grading accuracy and reduced energy consumption. We investigated flow field distributions in the proposed multi-product hydrocyclone by numerical simulation and performed experimental tests to facilitate understanding effects of structural and functional parameters on separation performance of hydrocyclone. This study provides references for understanding of grading by hydrocyclones and development of multi-product grading instrument in terms of both theory and industrial applications.

The remainder of this paper is organized as follows: Section 2 describes the mathematical model of the hydrocyclone with double vortex finders. Section 3 presents the numerical simulation results. Section 4 presents the tests results. The conclusions are summarized in Section 5.

2. Mathematical Model of Turbulent Flow Field of Hydrocyclones

The flow characteristics and flow motion inside the hydrocyclone are computed with continuity equation and momentum equation. Because the cyclone is cylindrical, therefore, the flow motion equation inside the cyclone is applied using cylindrical coordinates.

The continuity equation is expressed as follows:

$$\frac{1}{r}\frac{\partial u_\theta}{\partial \theta} + \frac{\partial u_r}{\partial r} + \frac{\partial u_z}{\partial z} + \frac{u_r}{r} = 0 \tag{1}$$

The momentum equation is expressed as follows:

$$\rho\left(u_r\frac{\partial u_r}{\partial r} + u_z\frac{\partial u_r}{\partial z} - \frac{u\theta^2}{r}\right) = -\frac{\partial \vec{P}}{\partial r} + \mu\left(\frac{\partial^2 u_r}{\partial r^2} + \frac{1}{r}\frac{\partial u_r}{\partial r} + \frac{\partial^2 u_r}{\partial z^2} - \frac{u_r}{r^2}\right) \tag{2}$$

$$\rho\left(u_r\frac{\partial u_\theta}{\partial r} + u_z\frac{\partial u_\theta}{\partial z} + \frac{u_r u_\theta}{r}\right) = \mu\left(\frac{\partial^2 u_\theta}{\partial r^2} + \frac{1}{r}\frac{u_\theta}{\partial r} + \frac{\partial^2 u_\theta}{\partial z^2} - \frac{u_\theta}{r^2}\right) \tag{3}$$

$$\rho\left(u_r\frac{\partial u_z}{\partial r} + u_z\frac{\partial u_z}{\partial z}\right) = -\frac{\partial \vec{P}}{\partial z} + \mu\left(\frac{\partial^2 u_z}{\partial r^2} + \frac{1}{r}\frac{u_z}{\partial r} + \frac{\partial^2 u_z}{\partial z^2}\right) \tag{4}$$

Reynolds proposed the theory of time-averaged flow field, which divides the instantaneous motion parameters into two parts, namely, the time-averaged values and the fluctuating values [19]. Therefore, the equation of continuity and the Navier-Stokes equation are written as follows:

$$\frac{1}{r}\frac{\partial u_\theta}{\partial \theta} + \frac{\partial u_r}{\partial r} + \frac{\partial u_z}{\partial z} + \frac{u_r}{r} = 0 \tag{5}$$

$$\rho\left(u_r\frac{\partial u_r}{\partial r} + u_z\frac{\partial u_r}{\partial z} - \frac{u\theta^2}{r}\right) = -\frac{\partial \vec{P}}{\partial r} + \mu\left(\frac{\partial^2 u_r}{\partial r^2} + \frac{1}{r}\frac{\partial u_r}{\partial r} + \frac{\partial^2 u_r}{\partial z^2} - \frac{u_r}{r^2}\right) \\ -\rho\left(\frac{1}{r}\frac{\partial}{\partial r}\overline{r u'_r{}^2} + \frac{\partial}{\partial z}\overline{u'_r u'_z} + \frac{1}{r}\overline{u'_\theta{}^2}\right) \tag{6}$$

$$\rho\left(u_r\frac{\partial u_\theta}{\partial r} + u_z\frac{\partial u_\theta}{\partial z} + \frac{u_r u_\theta}{r}\right) = \mu\left(\frac{\partial^2 u_\theta}{\partial r^2} + \frac{1}{r}\frac{u_\theta}{\partial r} + \frac{\partial^2 u_\theta}{\partial z^2} - \frac{u_\theta}{r^2}\right) \\ -\rho\left(\frac{\partial}{\partial r}\overline{u'_r u'_\theta} + \frac{\partial}{\partial z}\overline{u'_\theta u'_z} + \frac{2}{r}\overline{u'_r u'_\theta}\right) \tag{7}$$

$$\rho\left(u_r\frac{\partial u_z}{\partial r} + u_z\frac{\partial u_z}{\partial z}\right) = -\frac{\partial \vec{P}}{\partial z} + \mu\left(\frac{\partial^2 u_z}{\partial r^2} + \frac{1}{r}\frac{u_z}{\partial r} + \frac{\partial^2 u_z}{\partial z^2}\right) \\ -\rho\left(\frac{\partial}{\partial r}\overline{u'_r u'_z} + \frac{\partial}{\partial z}\overline{u'_z{}^2}\right) \tag{8}$$

3. Numerical Simulation

3.1. Modelling of Four Product Hydrocyclone

The first stage hydrocyclone is designed to be a 150 mm diameter double overflow pipe and the second cyclone was a conventional column cone 75 mm in diameter. The overflow of the first stage hydrocyclone was split into internal and external overflow paths. The external overflow, which is separated from the internal overflow by the overflow cap, is further graded in the second stage hydrocyclone. Figure 6 shows the structure model of the proposed hydrocyclone and Figure 7 shows the mesh. Then the mesh grid which determines the accuracy and the converging efficiency of numerical simulations that were developed for the model using meshing software ICEM. The fluid zone is divided into various hexahedron or tetrahedron meshes. The fitting degree is directly related to the mesh quantity. Mesh grids can be categorized as structural and non-structural grids. In structural grids [19,20] the mesh of each node in a specific zone is consistent with that of its adjacent node in this zone. These grids have advantages in boundary fitting, model accuracy, and data structure. However, structural grids are not well applicable for complicated flow field models. On the contrary, non-structural grids are applicable for any flow field model [21,22], although the calculation is usually time consuming and the accuracy is limited. In flow field models of four product hydrocyclones, the hexahedron structural grid was applied on a hydrocyclone column, while a non-structural grid was applied for feeding inlets and boundaries.

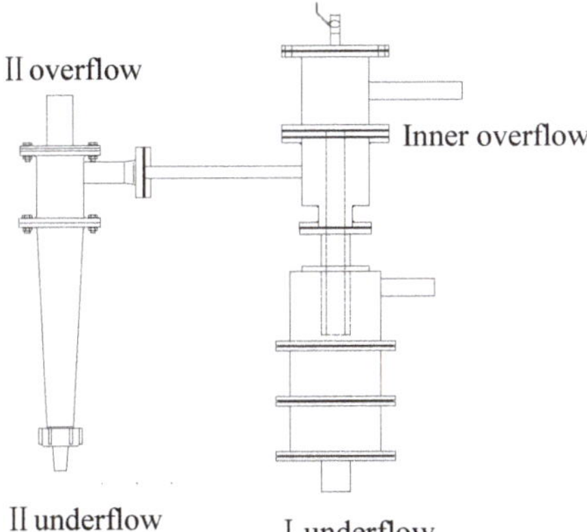

Figure 6. Structural representation of four product cyclone.

Figure 7. The meshing of four-products hydrocyclone.

Figure 8 represents the quality of the grid. As shown in Figure 8, the mesh quality whose critical level for simulations is 0.2, exceeds 0.35, indicating high quality of the mesh grid.

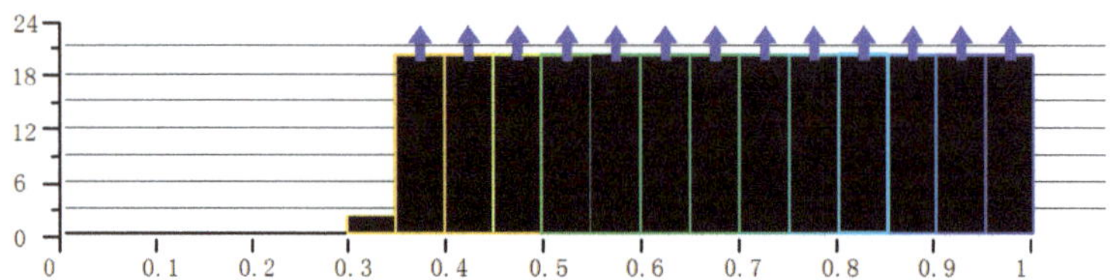

Figure 8. The mesh quality of cyclone.

3.2. Numerical Simulation

3.2.1. Selection of Turbulence Model and Multi-Phase Flow Model

In the Reynolds Stress equation Model (RSM) components of each stress are obtained by solving the Reynolds Stress equation. The RSM was selected as the turbulence model in this study [23–29].

In the multi-phase flow model, the VOF (Volume of Fluid) model is a simplified Euler-Euler model in which a momentum equation is solved and volume fraction of each fluid flowing through the computational domain is processed. The VOF model was selected to capture the liquid/gas interface in a hydrocyclone.

3.2.2. Boundary Condition and Calculation Scheme

Inlet boundary condition: the inlet velocity is defined as the inlet boundary condition. With the inlet velocity to be 2.5 m/s the turbulent intensity and hydraulic diameter are 3.8% and 36 mm respectively.

Outlet boundary condition: the outlet pressure is defined as the outlet boundary condition. For hydrocyclone, the outlet is exposed to atmosphere and the relative pressure is 0. Turbulence intensities and hydraulic diameters obtained are summarized in Table 1.

Table 1. Turbulence intensities and hydraulic diameters.

Parameters	I Underflow	II Underflow	Inner Overflow	II Overflow
Turbulence intensities/%	4.3	4.7	4.5	4.3
hydraulic diameter/mm	20	8	25	20

Numerical simulation and discrete scheme: numerical simulation was performed using the SIMPLE algorithm (pressure-velocity coupling). The discrete scheme for pressure was set to be PRESTO!, the discrete terms in momentum equation was set to be in First Order Upwind scheme and then turned into QUICK scheme after convergence. Discrete terms in all other functions were set to be in the First Order Upwind scheme.

3.2.3. Simulation and Results Analysis

- Effects of the first section cone angle on the flow fields in the first section hydrocyclone

Figure 9 summarizes pressures in the first section hydrocyclone at cone angles of 60°, 90°, 120°, and 180°.

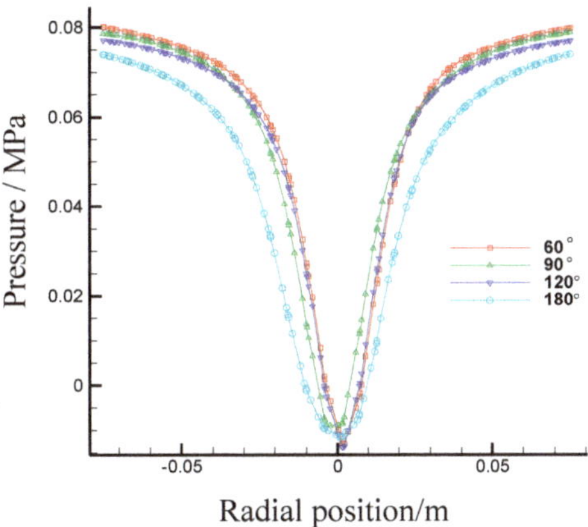

Figure 9. Pressure distribution of different cone angle hydrocyclone.

As observed, pressure in the hydrocyclone decreased from sidewall towards the core part and was centrosymmetrically distributed. As a result, a negative pressure zone was observed in the core part. The pressure in the first section hydrocyclone decreased as the first section cone angle increased.

As shown in Figure 10, the flow field velocity is affected by the cone angle: the tangential velocity of flow fields in the hydrocyclone decreased as the cone angle increased. This can be attributed to space increase in the hydrocyclone due to cone angle increase at fixed hydrocyclone height.

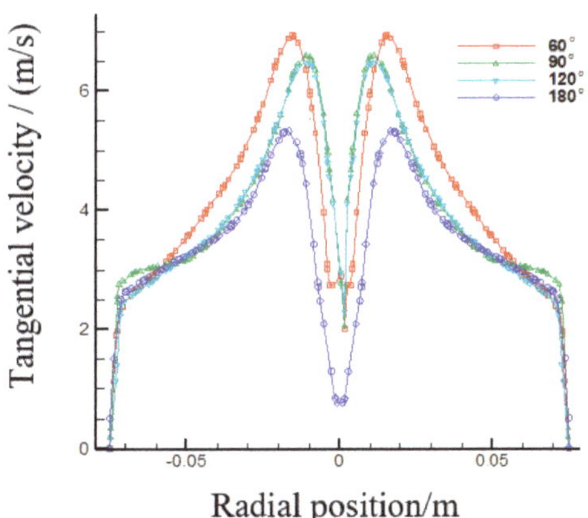

Figure 10. Tangential velocity distribution of different cone angle cyclone.

- Effects of the first section internal overflow pipe diameter on flow fields in the first section hydrocyclone

Figure 11 shows the internal pressure distributions in the first section internal overflow pipe in diameters of 15 mm, 20 mm, 25 mm, and 30 mm. The pressure was maximized (feeding pressure = 0.08~0.1 MPa) on the hydrocyclone sidewall and degraded along the radial direction. As a result, a negative pressure zone was observed in the core part. The pressure decreased as the internal overflow pipe diameter increased.

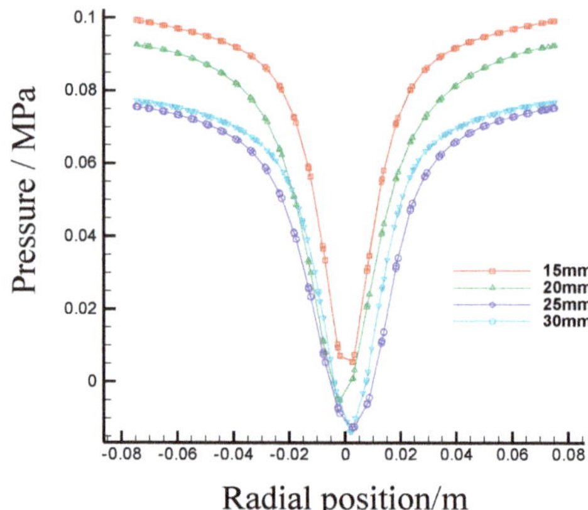

Figure 11. Pressure distribution of hydrocyclone with different inner vortex finder.

As shown in Figure 12, four curves correspond to tangential velocities of flow fields in the first section hydrocyclone at internal overflow pipe diameters of 15 mm, 20 mm, 25 mm, and 30 mm respectively. The tangential velocities increase along with the overflow outlet diameter increase. In addition, the location of the maximum tangential velocity shifted outwards along the radial direction as the internal overflow pipe diameter increased. This can be attributed to the outward shift of trajectory surface of maximum tangential velocity as a result of increase of the overflow pipe diameter.

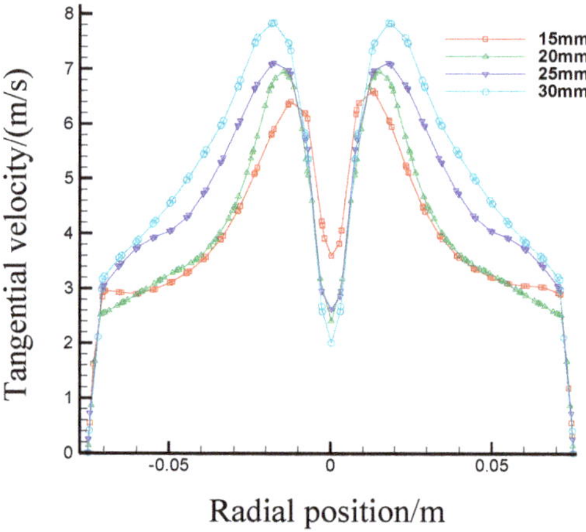

Figure 12. Velocity distribution of the hydrocyclone with a different inner vortex finder.

- Effects of the first section internal overflow pipe insertion depth on flow fields in the first section hydrocyclone

The internal overflow pipe insertion depth refers to the relative location of internal overflow pipe and external overflow pipe. The effects of insertion depth will be discussed in three situations. In the first case the internal overflow pipe is shorter than the external one and the internal overflow pipe outlet is located inside the external overflow pipe (insertion depth of internal overflow pipe = −50 mm). In the second case the internal overflow pipe is aligned with the external overflow pipe and the insertion depth of the internal overflow pipe is 0. In the third case the internal overflow pipe is longer than the external overflow pipe and the insertion depth of the external overflow pipe is either 50 mm or 100 mm, as shown in Figure 13.

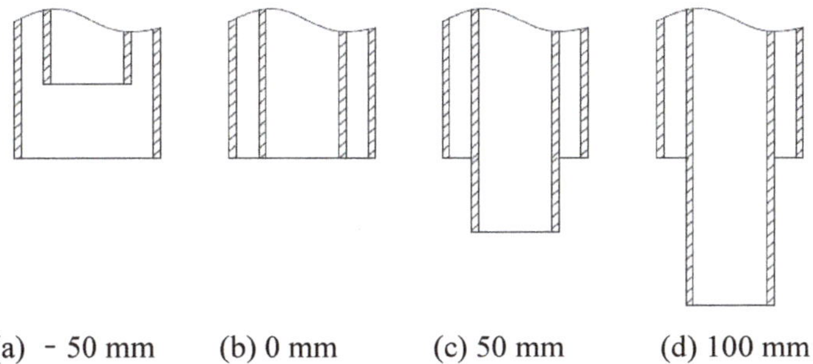

Figure 13. Insertion depth of inner vortex.

Figure 14 shows pressure distribution obtained by numerical simulation. Obviously, the effect of overflow pipe insertion depth is insignificant if the values of pressure at four different overflow pipe insertion depths are consistent.

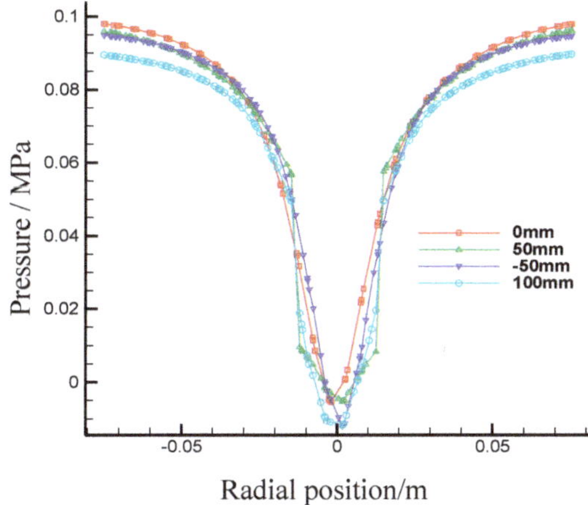

Figure 14. Pressure distribution inside cyclone with different insertion depth of inner vortex finder.

Figure 15 shows the tangential velocity of flow fields in the hydrocyclone vs. the insertion depth of the internal overflow pipe. One can see that the tangential velocity is maximized at 0 insertion depth from the sidewall towards the trajectory surface of the maximum tangential velocity. As this zone is the main area of separation in the hydrocyclone, the separation efficiency is improved by an increase of tangential velocity. The tangential velocity decreases from −50 mm and 50 mm and is minimized at 100 mm insertion depth. Also, the insertion depth affects axial velocity of flow fields and the secondary vortex is observed.

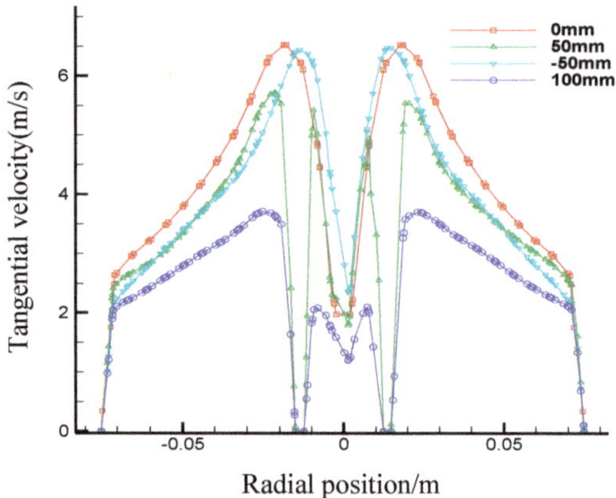

Figure 15. Tangential velocity inside cyclone with different insertion depth of inner vortex finder.

Figure 16 shows the distribution of axial velocities of flow fields in the first section hydrocyclone. We found that the axial velocity of fluids greatly varied and the distributions of flow fields are asymmetric. The axial velocity passed the zero point several times (changes of axial speed direction), resulting in a secondary vortex which facilitates grading but also leads to issues like an energy consumption gain and flow field instability.

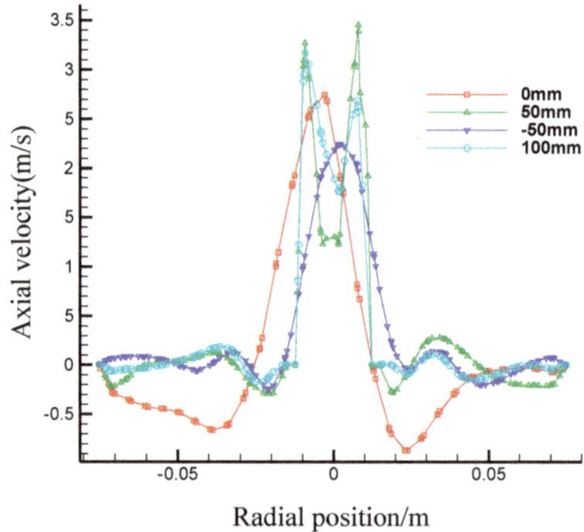

Figure 16. Axial velocity inside cyclone with different insertion depth of inner vortex finder.

- Effects of the first section underflow outlet diameter on flow fields in the first section hydrocyclone

Figure 17 shows pressure of flow field in the first section hydrocyclone vs. the first section underflow outlet diameter (12 mm, 16 mm, 20 mm, and 24 mm) by numerical simulations. As one can see, pressure decreases from the sidewall towards the core part along the radial direction and it decreases with an increase of the first section underflow outlet diameter d.

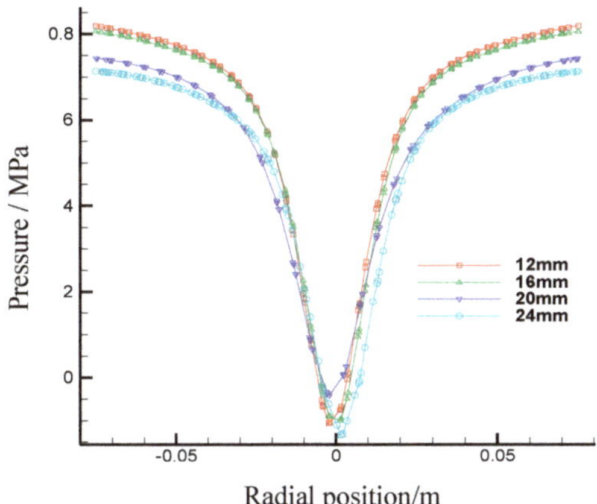

Figure 17. Pressure distribution inside cyclone with different I underflow pipe.

Figure 18 shows the tangential velocity of flow fields vs. the underflow outlet diameter. The tangential velocity of flow fields increases from the sidewall towards the core part along the radial direction. The maximum tangential velocity is observed at the interface between the inner and the outer swirls and it decreases after that. The tangential velocity of flow fields decreases with an increase of the underflow outlet diameter.

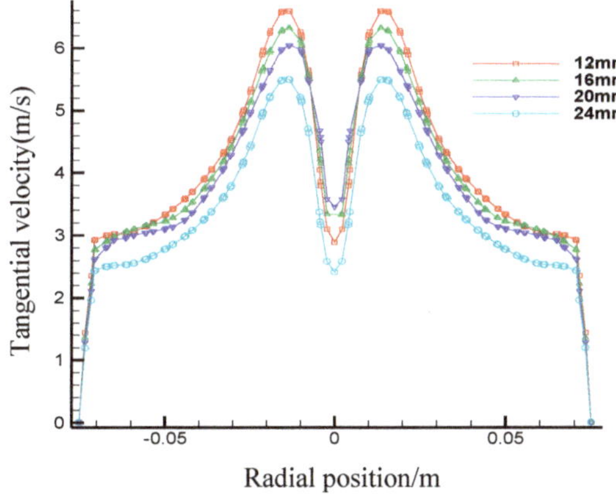

Figure 18. Tangential velocity inside cyclone with different I underflow pipe.

- Effects of the second section underflow outlet diameter on the first section hydrocyclone

Figure 19 shows pressure of flow field in the first section hydrocyclone obtained by numerical simulations vs. the second section underflow outlet diameter (6 mm, 8 mm, 10 mm, and 12 mm).

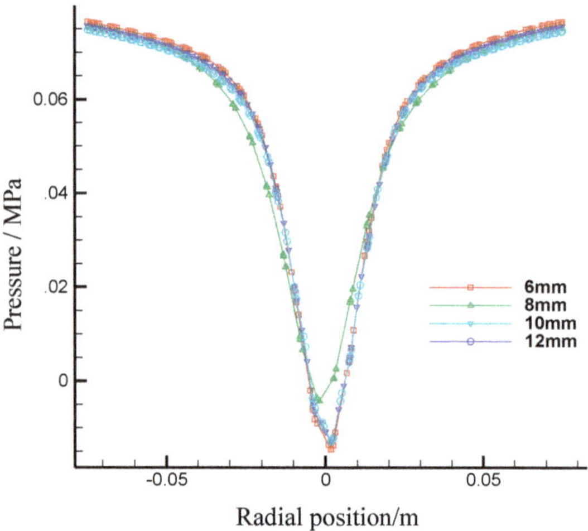

Figure 19. Pressure distribution inside cyclone with different II underflow pipe.

Figures 20 and 21 show distributions of tangential and axial velocities, respectively. It is evident that the second section underflow outlet diameter insignificantly affects distributions of pressure, tangential and axial velocities of flow field in the first section hydrocyclone.

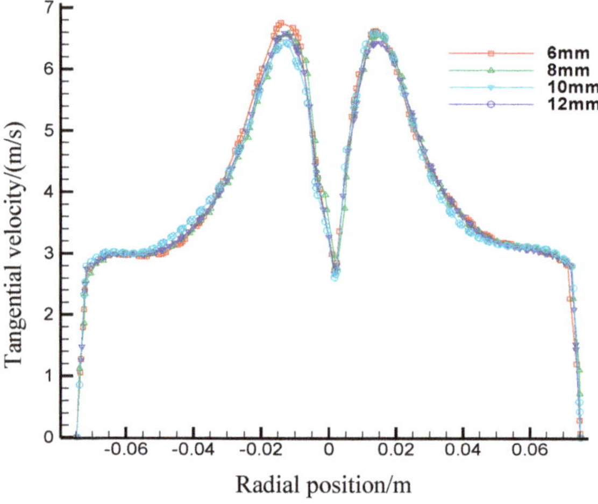

Figure 20. Tangential velocity inside cyclone with different II underflow pipe.

Figure 22 shows the radial velocity distribution curve in the first section hydrocyclone as a function of the second section underflow outlet diameter. In this case the radial velocity of the first section hydrocyclone increases along with the second section underflow outlet diameter increase. On one hand, radial velocity increase has a positive effect on grading efficiency; on the other hand, a radial velocity increase results in a rise in energy consumption.

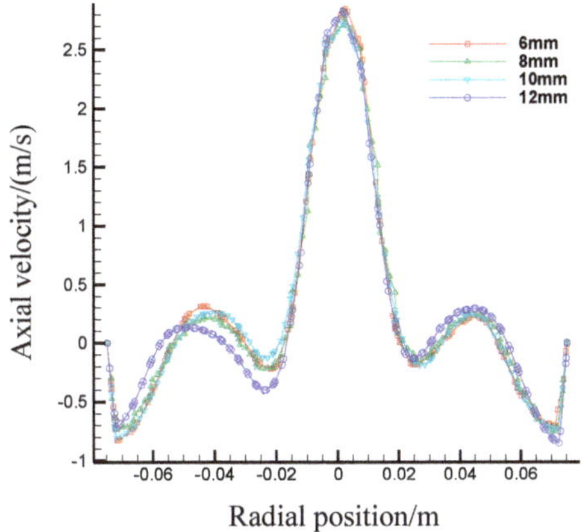

Figure 21. Axial velocity inside cyclone with different II underflow pipe.

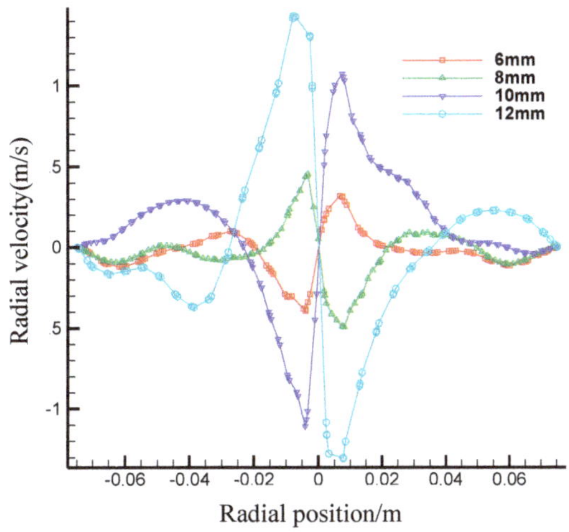

Figure 22. Radial velocity inside cyclone with different II underflow pipe.

4. Experimental Tests and Results

4.1. Procedure Design

For iron ore grading raw ore is grinded by the ball mill first. Then the ore is graded by conventional hydrocyclone: after monomer dissociation fine particles overflow goes to magnetic separator and iron concentrate is obtained; coarse particle underflow goes back to ball mill and hydrocyclone again for monomer dissociation. However, the conventional two product hydrocyclone provides fine particle overflow and coarse particle underflow in over-wide particle size ranges. Fine particles after monomer dissociation present in underflow, resulting in increased energy consumption and concentrate loss caused by over-grinding. We designed a technique for grading and recovery of iron ore fine particles using the proposed four product hydrocyclone, as shown in Figure 23.

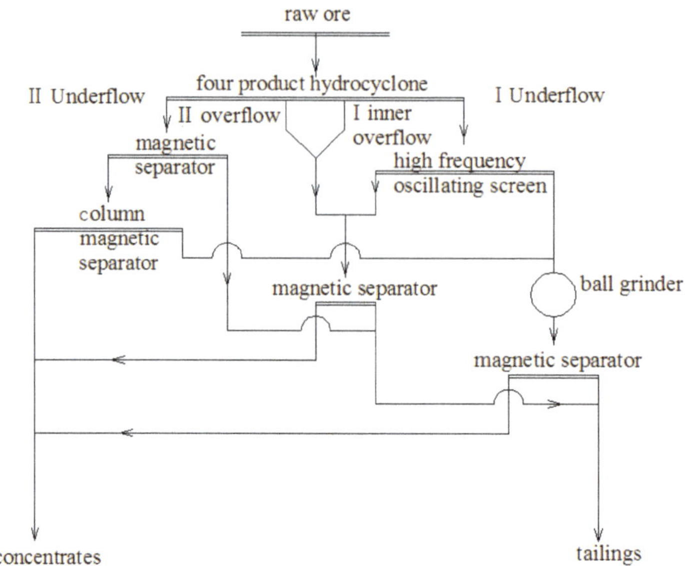

Figure 23. The process flow diagram of four-product cyclone.

After separation in the first stage hydrocyclone, the most coarse and dense ore particles joined in the underflow, while the finest particles joined in the internal overflow. The medium-size particles collect in the second stage hydrocyclone via the external overflow pipe and were further graded in the second stage overflow and the second stage underflow. Then, the first stage underflow of the hydrocyclone went to a high frequency oscillating screen. The mixture of under-sieve products from the internal overflow, and the second stage overflow went to a magnetic separator and concentrate was obtained.

4.2. Properties of Raw Ore

Table 2 summarizes properties of iron ore used in this study. As observed, particles under 38 μm reached 50.2% and the iron grade increases as the particle size decreases.

Table 2. The properties of ore.

Particle Size/mm	Yield/%	Cumulative Yield/%	Iron Grade/%	Iron Content/%	Distribution Rate/%	Cumulative Distribution Rate/%
+0.1	15.10	100	26.18	3.95	7.42	100
−0.1 + 0.076	10.72	84.90	37.93	4.07	7.65	92.58
−0.076 + 0.055	10.70	74.18	50.18	5.37	10.09	84.93
−0.055 + 0.043	13.28	63.48	58.51	7.77	14.60	74.84
−0.043 + 0.038	13.06	50.2	60.73	7.93	14.9	60.24
−0.038 + 0.030	6.53	37.14	61.07	3.99	7.50	45.34
−0.030	30.61	30.61	65.8	20.14	37.84	37.84
total	100.00		52.96	53.22	100.00	

4.3. Hydrocyclone Design

According to the iron sample analysis, particles under 30 μm are essentially concentrate. Therefore, the diameters of the first stage hydrocyclone and the second stage hydrocyclone are 50 mm and 25 mm, respectively. The structure is shown in Figure 6 and the structural parameters and operation parameters are summarized in Table 3.

Table 3. The optimal parameter based on overflow fineness.

Mass Concentration/%	Pressure/MPa	Angle of Cone/°	Diameter of I Underflow Pipe/mm	Diameter of II Underflow Pipe/mm	Diameter of Outer Overflow Pipe/mm	Diameter of Inner Overflow Pipe/mm
30	0.11	180	8	4	15	9

4.4. Result Analysis

The results of iron grading (Figure 24 shows the testing site of four product hydrocyclone) obtained by the procedure mentioned above are summarized in Table 4. The iron grade of concentrate obtained by the first section underflow is 66.24%. While combining the first stage internal overflow and the second stage overflow, the iron grade of concentrate is 62.53%. The iron grade of concentrate obtained from the second stage underflow is 61.14%. The iron grade of mixed concentrate was 65.08% (meeting the requirement of 65% for concentrate grade) and the concentrate recovery is 86.14%.

Figure 24. Four product hydrocyclone testing site.

Table 4. Industries operating results.

Results		Iron Grade/%	Iron Recovery Rate/%
I underflow	Concentrates after magnetic separation	66.24	49.22
	Tailings after magnetic separation	22.46	
merged overflow	Concentrates after magnetic separation	62.53	12.86
	Tailings after magnetic separation	28.71	
II underflow	Concentrates after magnetic separation	61.14	24.06
	Tailings after magnetic separation	24.42	
composite result	Concentrates after magnetic separation	65.08	86.14

5. Conclusions

To avoid the over-wide particle size range and a poor separation accuracy that does not meet requirements of the subsequent operations, a two-stage series, four product hydrocyclone is proposed.

The first stage hydrocyclone is designed to be a coaxial double overflow pipe: fine particles were obtained from the internal overflow and coarse particles were separated from the external overflow via the second section hydrocyclone. In this way, issues like over-wide particle size range and poor separation accuracy are avoided.

In experimental tests the proposed hydrocyclone was applied in iron recovery technology. Together with magnetic separator, oscillating screen, and ball mill, the proposed hydrocyclone can relieve the low iron concentrate recovery issue.

Author Contributions: Conceptualization, Y.Z. and P.L.; Methodology, Y.Z.; Software, Y.Z.; Validation, P.L. and L.J.; Formal Analysis, Y.Z.; Investigation, Y.Z. and L.J.; Resources, Y.Z.; Data Curation, X.Y.; Writing-Original Draft Preparation, Y.Z.; Writing-Review & Editing, Y.Z., P.L. and X.Y.

Funding: This research was funded by National Key R&D Program of China (2018YFC0604702), Natural Science Foundation of Shandong province (ZR2016EEM37) and key research and development project of Shangdong province(2017GSF216004).

Conflicts of Interest: The authors declare no conflict of interest.

References

1. Vakamalla, T.R.; Kumbhar, K.S.; Ravi Gujjula, N.M. Computational and experimental study of the effect of inclination on hydrocyclone performance. *Sep. Purif. Technol.* **2014**, *138*, 104–117. [CrossRef]
2. Svarovsky, L. *Hydrocyclones*; Technomic Publishing Co.: London, UK, 1984.
3. Banerjee, C.; Chaudhury, K.; Majumder, A.K.; Chakraborty, S. Swirling flow hydrodynamics in hydrocyclone. *Ind. Eng. Chem. Res.* **2015**, *54*, 522–528. [CrossRef]
4. Ni, L.; Tian, J.; Song, T.; Jong, Y.; Zhao, J. Optimizing Geometric Parameters in Hydrocyclones for Enhanced Separations: A Review and Perspective. *Sep. Purif. Technol.* **2018**, *48*, 30–51. [CrossRef]
5. Vega, D.; Brito-Parada, P.R.; Cilliers, J.J. Optimising small hydrocyclone design using 3D printing and CFD simulations. *Chem. Eng. J.* **2018**, *350*, 653–659. [CrossRef]
6. Hwang, K.; Chou, S. Designing vortex finder structure for improving the particle separation efficiency of a hydrocyclone. *Sep. Purif. Technol.* **2017**, *172*, 76–84. [CrossRef]
7. Jiang, L.; Liu, P.; Yang, X.; Zhang, Y. Short-Circuit Flow in Hydrocyclones with Arc-Shaped Vortex Finders. *Chem. Eng. Technol.* **2018**, *41*, 1783–1792. [CrossRef]
8. Yang, Y.; Wen, C. CFD modeling of particle behavior in supersonic flows with strong swirls for gas separation. *Sep. Purif. Technol.* **2017**, *174*, 22–28. [CrossRef]
9. Hsiao, T.; Huang, S.; Hsu, C.; Chen, C.; Chang, P. Effects of the geometric configuration on cyclone performance. *J. Aerosol Sci.* **2015**, *86*, 1–12. [CrossRef]
10. Krokhina, A.V.; Lvov, V.A.; Pavlikhin, G.P. A Probabilistic-Statistical Model of the Particle Classification Process in Small Hydrocyclone Classifiers. *Chem. Eng. Technol.* **2017**, *40*, 967–972. [CrossRef]
11. Liu, P.; Chu, L.; Wang, J.; Yu, Y. Enhancement of Hydrocyclone Classification Efficiency for Fine Particles by Introducing a Volute Chamber with a Pre-Sedimentation. *Chem. Eng. Technol.* **2008**, *31*, 474–478. [CrossRef]
12. Gonçalves, S.M.; Barrozo, M.A.S.; Vieira, L.G.M. Effects of solids concentration and underflow diameter on the performance of a newly designed hydrocyclone. *Chem. Eng. Technol.* **2017**, *40*. [CrossRef]
13. Surmen, A.; Avci, A.; Karamangil, M.I. Prediction of the maximum-efficiency cyclone length for a cyclone with a tangential entry. *Powder Technol.* **2011**, *207*, 1–8. [CrossRef]
14. Ahmed, M.M.; Ibrahim, G.A.; Farghaly, M.G. Performance of a three-product hydrocyclone. *Int. J. Miner. Process.* **2009**, *91*, 34–40. [CrossRef]
15. Obeng, D.P.; Morrell, S. The JK three-product cyclone-performance and potential applications. *Int. J. Miner. Process.* **2003**, *69*, 129–142. [CrossRef]
16. Obeng, D.P.; Morrell, S.; Napier-Munn, T.J. Application of central composite rotatable design to modeling the effect of some operating variables on the performance of the three-product cyclone. *Int. J. Miner. Process.* **2005**, *76*, 181–192. [CrossRef]
17. Mainza, A.; Narasimha, M.; Powell, M.S.; Holtham, P.N.; Brennan, M. Study of flow behavior in a three-product cyclone using computational fluid dynamics. *Miner. Eng.* **2006**, *19*, 1048–1058. [CrossRef]

18. Mainza, A.; Powell, M.S.; Knopjes, B. Differential classification of dense material in a three-product cyclone. *Miner. Eng.* **2004**, *17*, 573–579. [CrossRef]
19. Bhasker, C. Flow simulation in industrial cyclone separator. *Adv. Eng. Soft.* **2010**, *41*, 220–228. [CrossRef]
20. Chen, J.; Chu, K.; Zou, R.; Yu, A.B.; Vince, A.; Barnett, G.D.; Barnett, P.J. Systematic study of the effect of particle density distribution on the flow and performance of a dense medium cyclone. *Powder Technol.* **2017**, *314*, 510–523. [CrossRef]
21. Vakamalla, T.R.; Korprolu, V.B.R.; Arugonda, R.; Mangadoddy, A. Development of novel hydrocyclone designs for improved fines classification using multiphase CFD model. *Sep. Purif. Technol.* **2017**, *175*, 481–497. [CrossRef]
22. Slack, M.D.; Prasad, R.O.; Bakker, A.; Boysan, F. Advances in cyclone modeling using unstructured grids. *Chem. Eng. Res.* **2000**, *78*, 1098–1104. [CrossRef]
23. Wang, S.; Luo, K.; Hu, C.; Fan, J. CFD-DEM study of the effect of cyclone arrangements on the gas-solid flow dynamics in the full-loop circulating fluidized bed. *Chem. Eng. Sci.* **2017**, *172*, 199–215. [CrossRef]
24. Siadaty, M.; Kheradmand, S.; Ghadiri, F. Improvement of the cyclone separation efficiency with a magnetic field. *J. Aerosol Sci.* **2017**, *114*, 219–232. [CrossRef]
25. Parvaz, F.; Hosseini, S.H.; Ahmadi, G.; Elsayed, K. Impacts of the vortex finder eccentricity on the flow pattern and performance of a gas cyclone. *Sep. Purif. Technol.* **2017**, *187*, 1–13. [CrossRef]
26. Cui, B.; Zhang, C.; Wei, D.; Lu, S.; Feng, Y. Effects of feed size distribution on separation performance of hydrocyclones with differdent vortex finder diameters. *Powder Technol.* **2017**, *322*, 114–123. [CrossRef]
27. Patra, G.; Velpuri, B.; Chakraborty, S.; Meikap, B.C. Performance evaluation of a hydrocyclone with a spiral rib for separation of particles. *Adv. Powder Technol.* **2017**, *28*, 3222–3232. [CrossRef]
28. Zhang, Y.; Cai, P.; Jiang, F.; Dong, K.; Jiang, Y.; Wang, B. Understanding the separation of particles in a hydrocyclone by force analysis. *Powder Technol.* **2017**, *322*, 471–489. [CrossRef]
29. Ji, L.; Kuang, S.; Qi, Z.; Wang, Y.; Chen, J.; Yu, A. Computational analysis and optimization of hydrocyclone size to mitigate adverse effect of particle density. *Sep. Purif. Technol.* **2017**, *174*, 251–263. [CrossRef]

Article

The Effect of Inlet Velocity on the Separation Performance of a Two-Stage Hydrocyclone

Lanyue Jiang, Peikun Liu *, Yuekan Zhang, Xinghua Yang and Hui Wang

College of Mechanical & Electronic Engineering, Shandong University of Science and Technology, Qingdao 266590, China; jianglanyue5@163.com (L.J.); zhangyk2007@163.com (Y.Z.); yxh19781025@126.com (X.Y.); wang_hui_1992@163.com (H.W.)
* Correspondence: lpk710128@163.com; Tel.: +86-0532-8605-7176

Received: 28 February 2019; Accepted: 26 March 2019; Published: 30 March 2019

Abstract: The "entrainment of coarse particles in overflow" and the "entrainment of fine particles in underflow" are two inevitable phenomena in the hydrocyclone separation process, which can result in a wide product size distribution that does not meet the requirement of a precise classification. Hence, this study proposed a two-stage (TS) hydrocyclone, and the effects of the inlet velocity on the TS hydrocyclone were investigated using computational fluid dynamics (CFD). More specifically, the influences of the first-stage inlet velocity on the second-stage swirling flow field and the separation performance were studied. In addition, the particle size distribution of the product was analyzed. It was found that the first-stage overflow contained few coarse particles above 40 μm and that the second-stage underflow contained few fine particles. The second-stage underflow was free of particles smaller than 10 μm and almost free of particles smaller than 20 μm. The underflow product contained few fine particles. Moreover, the median particle size of the second-stage overflow product was similar to that of the feed. Inspired by this observation, we propose to recycle the second-stage overflow to the feed for re-classification and to use only the first-stage overflow and the second-stage underflow as products. In this way, fine particle products free of coarse particle entrainment, and coarse particle products free of fine particle entrainment can be obtained, achieving the goal of precise classification.

Keywords: two-stage (TS) hydrocyclone; computational fluid dynamics; separation performance; particle size; inlet velocity

---

1. Introduction

A hydrocyclone is an effective device for multiphase separation, utilizing the principle of centrifugal sedimentation. It is widely used in grading, desliming, concentration, clarification, and sorting because of its simple design, convenient operation, high capacity, and high separation efficiency [1–5]. Precise size classification is one of the main means to improve the separation efficiency and sharpness of the beneficiation process. Currently, the hydrocyclone is the main device for particle size classification [6–8]. Two products, overflow and underflow, can be obtained via a single separation step in a conventional hydrocyclone. Particles with sizes smaller than the cut size fall into the overflow, and particles with sizes larger than the cut size fall into the underflow. However, the "entrainment of coarse particles in the overflow" and the "entrainment of fine particles in the underflow" are two inevitable phenomena that occur because of the complexity of the hydrocyclone's internal flow field, and these phenomena can result in a wide product size distribution that does not meet the requirement of precise classification [9–12]. Connecting multiple hydrocyclones in series is one way to solve the problem of wide product size distribution and obtain a series of products with narrow size distributions. Although this method can produce multiple products with different particle sizes, it requires additional feed pumps and pipelines, which causes problems such as a long process flow, a large equipment investment, and high energy consumption.

Although a hydrocyclone has a simple structure, it has a complicated internal flow field distribution and separation mechanism. It is difficult to design a hydrocyclone structure that can be generally applied to various industrial fields. Hence, it is necessary to develop hydrocyclones with different structures for different separation processes and different applications. For many years, researchers worldwide studied the separation performance of hydrocyclones with different structures. The work of structural improvement of hydrocyclones focused on how to improve the classification efficiency and to achieve multi-functionality [13–17]. Restarick [18,19], Zhang [20], and others designed a two-stage, three-product heavy–medium cyclone that consisted of a primary cylindrical cyclone and a secondary cylindrical/conical cyclone. Three products can be obtained in a single separation step. In the primary cylindrical cyclone, particles were mainly classified by size so that high-ash fine coal can be discharged through the overflow. In the secondary cylindrical/conical cyclone, particles were mainly classified by density, which can effectively separate the clean coal and the gangue. Obeng [21,22] developed a Julius Kruttschnitt (JK) three-product cyclone. In their design, a vortex finder with a smaller inner diameter was inserted coaxially into the conventional vortex finder. This type of hydrocyclone can produce three products in a single separation step: a fine overflow stream, a middling overflow stream, and a coarse underflow stream. This study also investigated the effect of the insertion depth of the second vortex finder on the separation performance of the cyclone. Ahmed [23] designed a three-product, solid–liquid separation hydrocyclone with one overflow and two underflows by adding a tangential output opening in the middle of the conical section on the same side as the feed inlet opening. Krokhina [24], Farghal [25], Golyk [26], and Dueck [27] installed a water-injection system in the conical section of a conventional cylindrical/conical cyclone near the apex so that the material was subjected to a secondary classification process to meet the classification requirements for particles of different sizes. In this way, the fine particle fraction in the underflow can be reduced effectively. The separation efficiency and separation sharpness can both be improved. Although these studies improved the performance of the cyclone to a certain extent, many aspects of the cyclone still need to be explored. Because of the influences of the number of outlets and the cyclone structure, the pressure drop and separation performance of a multi-product cyclone during operation are significantly different from those of a conventional single-stage cyclone. For two-stage series cyclones, especially the underflow-fed series cyclones, the inlet velocity of the first stage has a great influence on the second stage. If the inlet velocity of the first stage is too small, it is difficult to form a sufficiently strong centrifugal force field in the second stage after the energy loss in the first stage. Therefore, it is necessary to study the influence of inlet velocity on the flow field and the separation performance of two-stage series cyclones.

In recent years, with the development of computer technology, computational fluid dynamics (CFD) became an effective technique for fluid research and it is widely used in the study of the flow field and the separation characteristics of cyclones [28–30]. Delgadillo and Rajamina [31] designed and simulated six 75-mm hydrocyclones with various cylindrical geometries and cone angles. They studied the variation law of the air core using the Reynolds stress model (RSM) and the large eddy simulation (LES) model coupled with the volume of fluid (VOF) model, and they studied the particle trajectory law with the aid of the discrete phase model (DPM). Compared with the standard hydrocyclone design, their novel designs had smaller cut sizes and better separation efficiencies. Narasimha [32] found that both the RSM and LES models could accurately predict the velocity field in hydrocyclones. Brennan [33] predicted the tangential velocity, axial velocity, and radial pressure profiles in a hydrocyclone and compared the simulation results with the laser Doppler velocimetry (LDV) test results on a 75-mm hydrocyclone. The results showed that the velocity and pressure profiles of the simulation results were consistent with the test results in the upper conical section and the cylindrical section. Slack [34] simulated the turbulence field in a 205-mm hydrocyclone using the RSM and LES models. The obtained axial and tangential velocity profiles were compared with the LDV test data. The RSM turbulence model-generated predictions agreed well with the LDV test data and had lower requirements for computational cost and mesh quality than the LES model. Conversely, the LES model required refined meshes and smaller time steps, making it much more computationally expensive than the RSM model. Therefore, the RSM model is generally preferred for simulations, and its accuracy was successfully proven in recent studies. To study multiphase flow in a hydrocyclone, the VOF model is widely adopted to investigate the air core inside a hydrocyclone. A particle image velocimetry (PIV) test and other flow field tests verified that the VOF model can accurately predict the distribution characteristics of the gas–liquid two-phase flow in the hydrocyclones [35]. Lagrangian particle tracking (LPT) and the two-fluid model (TFM) are widely used to study particle separation. The former traces the motion of a single particle but ignores the influence of inter-particle interactions and the reactions of particles on the fluid [36]. The TFM model, treating both liquid and solid phases as interpenetrating continuous media, incorporates the interaction between particles and between the particles and the fluids. This model is increasingly used to study the effect of concentration on the separation performance of hydrocyclones [37]. In this study, the effects of inlet velocity on the TS hydrocyclone were investigated using CFD. More specifically, the influences of the first-stage inlet velocity on the second-stage swirling flow field and the separation performance were studied.

2. Numerical Methods

2.1. Structure Geometry

An underflow-fed series TS hydrocyclone was designed to study the influence of inlet velocity on flow field and separation performance. The schematic diagram of the TS hydrocyclone is shown in Figure 1, and its specific structural parameters are given in Table 1. When slurry tangentially enters the first stage of the hydrocyclone at a certain speed, the fluid is forced to spiral along the hydrocyclone wall at a high speed, and a centrifugal force field is formed. The centrifugal force that fine particles are subjected to cannot overcome the fluid resistance. Thus, fine particles follow the fluid toward the center of the hydrocyclone and then exit the vortex finder to form a first-stage overflow stream. Coarse particles and some fine particles can overcome the fluid resistance and move outward to the sidewalls, before continuing to spiral downward under the influence of the subsequent fluid flow until they exit the tangential opening of the first stage and enter the second stage for secondary separation. Fluid that enters the second stage of the hydrocyclone is still able to form a centrifugal force field under the residual pressure. When particles enter the second stage, coarse particles are quickly separated and discharged from the underflow to form a second-stage underflow stream. The remaining fine particles are discharged from the vortex finder to form a second-stage overflow. Three grades of products can be obtained by a single TS hydrocyclone separation operation.

Table 1. Structural parameters of hydrocyclones.

Structural Parameters	Size
Diameter of first cylinder section D1 (mm)	75
Diameter of second cylinder section D2 (mm)	50
Height of first-stage cylinder section H1 (mm)	150
Height of first-stage body H2 (mm)	180
Height of first-stage vortex finder Ho1 (mm)	50
Diameter of inlet a × b (mm)	15 × 20
Height of second stage body H3 (mm)	281
Height of second-stage cylinder section H4 (mm)	85
Height of second-stage cone section H5 (mm)	150
Height of second-stage outer vortex finder H6 (mm)	30
Height of second-stage inner vortex finder Ho2 (mm)	50
Diameter of apex Du2 (mm)	7
Diameter of second vortex finder Do2 (mm)	15
Diameter of first vortex finder Do1 (mm)	25

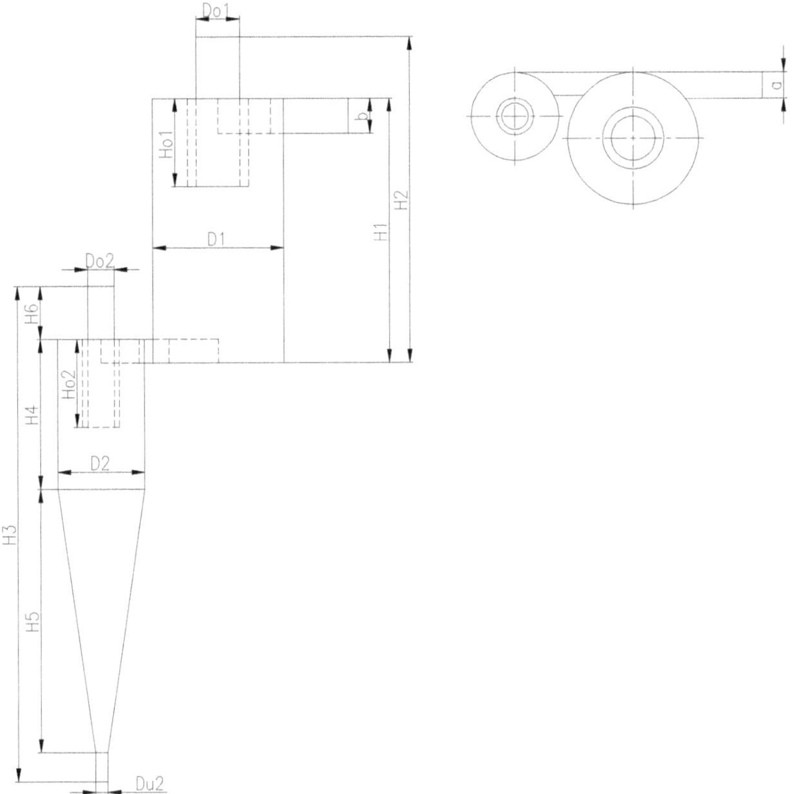

Figure 1. Two-stage (TS) hydrocyclone.

2.2. Model Description

(1) RSM Model

The RSM model was used, and its transport equation is as follows:

$$\frac{\partial\left(\rho\overline{u_i'u_j'}\right)}{\partial t} + \frac{\partial\left(\rho u_k\overline{u_i'u_j'}\right)}{\partial x_k} = D_{T,ij} + D_{L,ij} + P_{ij} + G_{ij} + \Phi_{ij} + \varepsilon_{ij} + F_{ij}, \tag{1}$$

where $D_{T,ij}$ is the turbulent energy diffusion, $D_{L,ij}$ is the molecular viscous diffusion, P_{ij} is the shear stress generation, G_{ij} is the buoyancy generation, ϕ_{ij} is the pressure strain, ε_{ij} is the viscous discrete term, and F_{ij} is the system rotation generation. The equations for each are as follows:

$$D_{T,ij} = -\frac{\partial}{\partial x_k}\left[\rho\overline{u_i'u_j'u_k'} + \overline{p'u_i'}\delta_{kj} + \overline{p'u_j'}\delta_{jk}\right]; \tag{2}$$

$$D_{L,ij} = \frac{\partial}{\partial x_k}\left[\mu\frac{\partial}{\partial x_k}\left(\overline{u_i'u_j'}\right)\right]; \tag{3}$$

$$P_{ij} = -\rho\left(\overline{u_i'u_k'}\frac{\partial u_j}{\partial x_k} + \overline{u_j'u_k'}\frac{\partial u_i}{\partial x_k}\right); \tag{4}$$

$$G_{ij} = -\rho\beta\left(g_i\overline{u_j'\theta} + g_j\overline{u_i'\theta}\right); \tag{5}$$

$$\Phi_{ij} = p'\overline{\left(\frac{\partial u'_i}{\partial x_j} + \frac{\partial u'_j}{\partial x_i}\right)};\tag{6}$$

$$\varepsilon_{ij} = -2\mu\overline{\frac{\partial u'_i}{\partial x_k}\frac{\partial u'_j}{\partial x_k}};\tag{7}$$

$$F_{ij} = -2\rho\Omega_k\left(\overline{u'_j u'_m}e_{ikm} + \overline{u'_i u'_m}e_{jkm}\right).\tag{8}$$

In this paper, the internal flow field of the hydrocyclone was numerically simulated using ANSYS Fluent software (ANSYS Inc, Canonsburg, PA, USA), and the turbulent energy diffusion $D_{T,ij}$, stress–strain ϕ_{ij}, buoyancy generation G_{ij}, and dissipative tensor ε_{ij} used in Fluent are as follows:

$$D_{T,ij} = -\frac{\partial}{\partial x_k}\left[\frac{\mu}{\sigma_k}\frac{\partial \overline{u_i u_j}}{\partial x_k}\right];\tag{9}$$

$$\Phi_{ij} = \phi_{ij,1} + \phi_{ij,2} + \phi_{ij}^w;\tag{10}$$

$$G_{ij} = \beta\frac{\mu_t}{Prt}\left(g_i\frac{\partial T}{\partial x_j} + g_j\frac{\partial T}{\partial x_i}\right);\tag{11}$$

$$\varepsilon_{ij} = -\frac{2}{3}\delta_{ij}(\rho\varepsilon + Y_M).\tag{12}$$

In the equations, $\mu_t = \rho C_\mu\left(k^2/\varepsilon\right)$, $\sigma_k = 0.82$, $\phi_{ij,1}$ is the slow term, $\phi_{ij,2}$ is the fast term, ϕ_{ij}^w is the wall reflection, and $Y_M = 2\rho\varepsilon M_t^2$, where M_t is the Mach number.

(2) TFM Model

The mass conservation equation is as follows:

$$\frac{\partial}{\partial t}(\rho_m) + \nabla\cdot\left(\rho_m\vec{v}_m\right) = \dot{m},\tag{13}$$

where ρ_m is the mixed phase density, $\dot{m}$ is the mass transfer, and $\vec{v}_m$ is the average velocity of the mixed phase.

The average velocity of the mixed phase is as follows:

$$\vec{v}_m = \frac{\sum\limits_{k=1}^{n}\alpha_k\rho_k\vec{v}_k}{\rho_m},\tag{14}$$

where α_k is the volume fraction of the kth phase, and $\vec{v}_k$ is the velocity of the kth phase.

The momentum conservation equation is as follows:

$$\frac{\partial}{\partial t}\left(\rho_m\vec{v}_m\right) + \nabla\cdot\left(\rho_m\vec{v}_m\vec{v}_m\right) = -\nabla\cdot\vec{p} + \nabla\cdot\left[\mu_m\left(\nabla\vec{v}_m + \nabla\vec{v}_m^T\right)\right] + \rho_m\vec{g} + \vec{F} + \nabla\cdot\left(\sum_{k=1}^{n}\alpha_k\rho_k\vec{v}_k^r\vec{v}_k^r\right),\tag{15}$$

where n is the total number of phases, μ_m is the viscosity of the mixed phase, $\vec{p}$ is the pressure, ρ_k is the density of the kth phase, $\vec{F}$ is the volume force, $\vec{v}_k^r$ is the relative sliding velocity between the kth phase and the mixed phase, and $\vec{v}_k^r = \vec{v}_k - \vec{v}_m$.

The volume fraction equation for the kth phase is as follows:

$$\frac{\partial}{\partial t}(\alpha_k\rho_k) + \nabla\cdot\left(\alpha_k\rho_k\vec{v}_m\right) = -\nabla\cdot\left(\alpha_k\rho_k\vec{v}_k^r\right).\tag{16}$$

2.3. Simulation Conditions

To study the influence of inlet velocity on the flow field and the separation performance of the TS hydrocyclone, a Cartesian coordinate system was constructed, with its origin defined as the center of the second-stage apex. The x-axis points in the feed direction of the first stage of the hydrocyclone. The z-axis points in the vortex finder direction of the second stage of the hydrocyclone. The mathematical model of the hydrocyclone was established and meshed using block-structured hexahedral mesh generation techniques, as shown in Figure 2a; the total number of mesh elements was approximately 353,500. Mesh division is the most important step in the pre-processing of a numerical simulation. The number, quality, and type of mesh have great influence on the accuracy and speed of simulation calculation. Meanwhile, in order to ensure the calculating accuracy and improve the computational efficiency, mesh independence detection was carried out. Tangential velocity is an important performance index of hydrocyclone; thus, the influence of grid on simulation precision was evaluated by the change of tangential velocity, and the results are shown in Figure 2b. When the number of meshes is greater than 300,000, the tangential velocity of the hydrocyclone is basically unchanged with the increase in the number of meshes. The orthogonality of grid lines, the aspect ratio, and the connectivity of grids were auto-checked after the mesh creation.

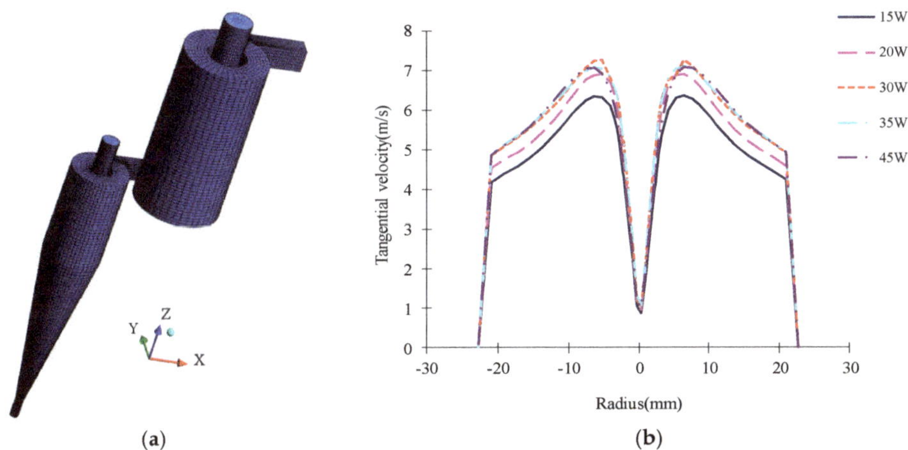

Figure 2. The mesh of TS hydrocyclone: (**a**) mesh; (**b**) mesh independence detection.

The RSM and TFM models were employed to simulate the internal flow field and to calculate the particle classification efficiency of the TS hydrocyclone. In all simulations, the "velocity-inlet" boundary condition was applied at the hydrocyclone inlet. The water and solid particles had the same speed, which was in the range of 3–6 m/s. The "pressure-outlet" boundary condition was applied at the outlet. The outlet pressure was set to standard atmospheric pressure. The solid particles used in the TFM model were SiO_2 particles with a density of 2650 kg/m^3. The particle size distribution is given in Table 2. The total solid volume fraction in the feed was 3.5%. A no-slip boundary condition was applied to the hydrocyclone walls. The SIMPLE algorithm was used for the pressure–velocity coupling. The PRESRO discretization scheme was used for the pressure equations, and the QUICK discretization scheme was used for other control equations. The time-averaged equilibrium of the flow rate of each phase at both the inlet and the outlet was used as the convergence criterion. The simulation time step was set to 1.0×10^{-4} s.

Table 2. The size distribution of particles in the feed.

Size Interval (μm)	Mean Size (μm)	Yield (%)	Volume Fraction (%)
0–2.5	1	4.88	0.1708
2.5–7.5	5	7.21	0.2524
7.5–15	10	9.55	0.3343
15–25	20	15.33	0.5366
25–35	30	20.88	0.7308
35–45	40	16.59	0.5807
45–55	50	11.38	0.3983
55–65	60	8.31	0.2909
65–80	75	5.87	0.2055

3. Results and Discussion

3.1. Model Validation

It is critical to firstly validate the mathematical model with accurate test data before it is applied in numerical experiments. In 1998, Hsieh [38] obtained the velocity profile on the cross-section of a Φ75 hydrocyclone using LDV. This test result was used by many researchers to validate their simulation results. Therefore, in this work, we firstly compared the simulation results with Hsieh's experimental data. As shown in Figure 3, the calculated axial and tangential velocity profiles are in good agreement with the experimental data, which indicates that the selected mathematical model can effectively predict the velocity profiles in the hydrocyclones.

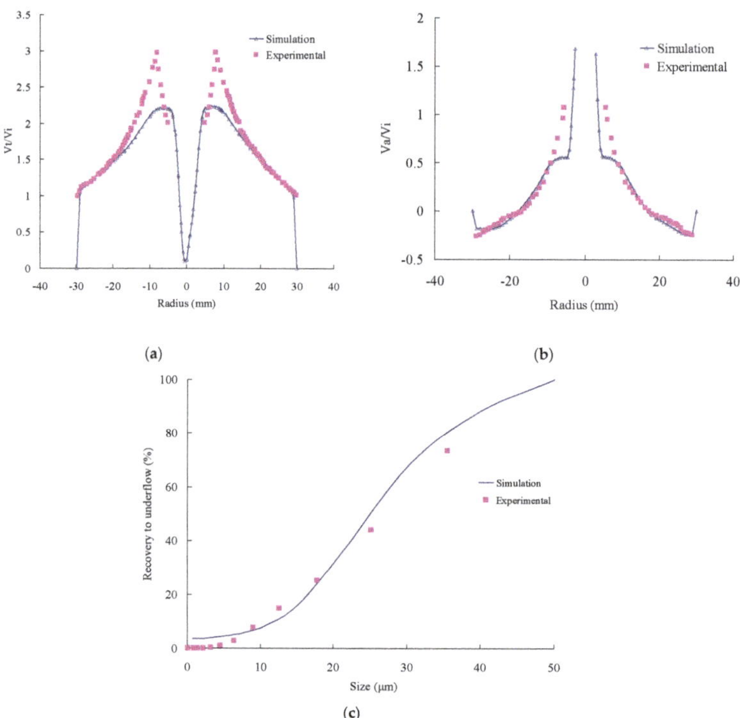

Figure 3. Comparisons between the simulation and the classic experiment: (**a**) tangential velocity (with Hsieh's results [38]); (**b**) axial velocity (with Hsieh's results); (**c**) comparison of separation behaviours (with Delgadillo's results [3]).

The TFM model was validated based on the experimental work of Delgadillo [3]. The solid content of the feed in Delgadillo's work was 10.47% by weight. The same particle size distribution of the feed was used in the present study as in Delgadillo's work. The classification efficiency was characterized by measuring the recoveries of particles with different size classes in the underflow. Figure 3c shows that the measured and simulated partition curves are in good agreement and that the overall trends are similar for the two approaches. Therefore, we conclude that the TFM model used in the present study can accurately predict the separation behavior of the hydrocyclone.

3.2. Flow Pattern

The inlet velocity directly affects the strength of the centrifugal force field in the hydrocyclone and is an important indicator of the hydrocyclone's performance. If the inlet velocity is too small, a sufficient centrifugal force field that is required for particle separation cannot be formed. If the inlet velocity is too large, the high flow rate shortens the residence time of the particles in the hydrocyclone, which may hinder the complete separation of the particles and unnecessarily increase the energy consumption. Therefore, it is necessary to analyze the influence of the inlet velocity on the separation performance of the hydrocyclone.

3.2.1. Pressure

Figure 4 shows the relationship between the pressure drop and the inlet velocity in the hydrocyclone. As the inlet velocity increases, the pressure drop in the hydrocyclone increases linearly. When the inlet velocity increases from 3 m/s to 6 m/s, the pressure drop of the first stage of the hydrocyclone increases from 27.81 kPa to 116.21 kPa, and the pressure drop of the second stage of the hydrocyclone increases from 26.14 kPa to 111.17 kPa. To ensure the basic separation performance of a hydrocyclone, its pressure drop generally needs to be greater than 50 kPa. For a TS hydrocyclone, sufficient pressure in the first stage of the hydrocyclone must be maintained to ensure the separation performance of the second stage of the hydrocyclone. Therefore, it is necessary to keep the inlet velocity greater than 4 m/s to maintain the performance of the TS hydrocyclone.

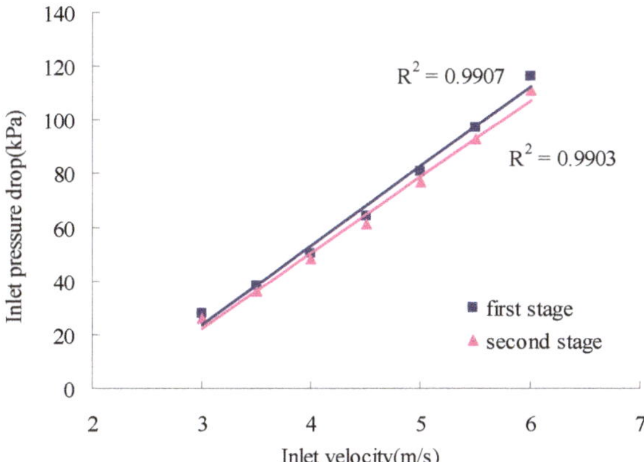

Figure 4. Relationship between pressure drop and inlet velocity.

The pressure distributions inside the TS hydrocyclone are shown in Figure 5, which shows that the pressure distributions in the first and second stages are essentially the same and are both in accordance with the pressure distributions of the semi-free vortex and the forced vortex. The pressure is axisymmetrically distributed with its highest value at the wall surface, and gradually decreases from the wall toward the axial center with decreasing radius. However, near the center of the hydrocyclone, the first stage is 5.5 mm away from the center (about 0.44 of the diameter of the vortex finder), and the second stage is 2.5 mm away from the center (about 0.33 of the diameter of the vortex finder), the pressure becomes negative, and the pressure at the axial center is also negative. This central negative pressure region is the cause of the air core at the center of the hydrocyclone. Figure 5 also shows that the radial pressure gradient gradually increases as the inlet velocity increases. To further analyze the radial pressure distribution, the radial pressure distributions were simulated at the cros- sections of $Z = 300$ mm (in the first stage of the hydrocyclone) and $Z = 150$ mm (in the second stage of the hydrocyclone). The results are shown in Figure 6.

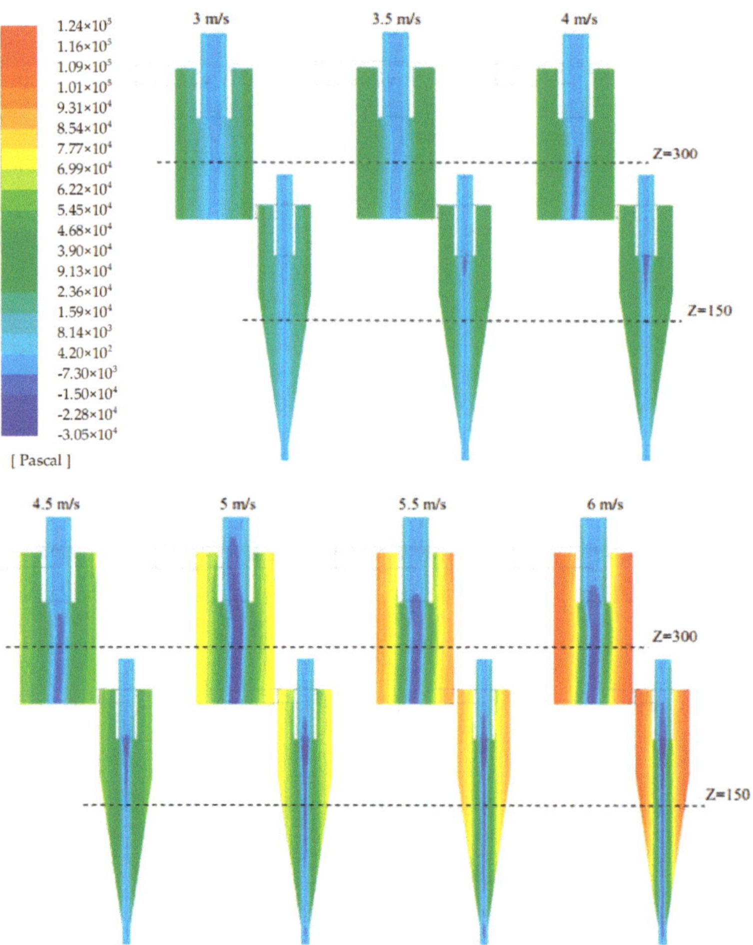

Figure 5. Pressure distribution contours.

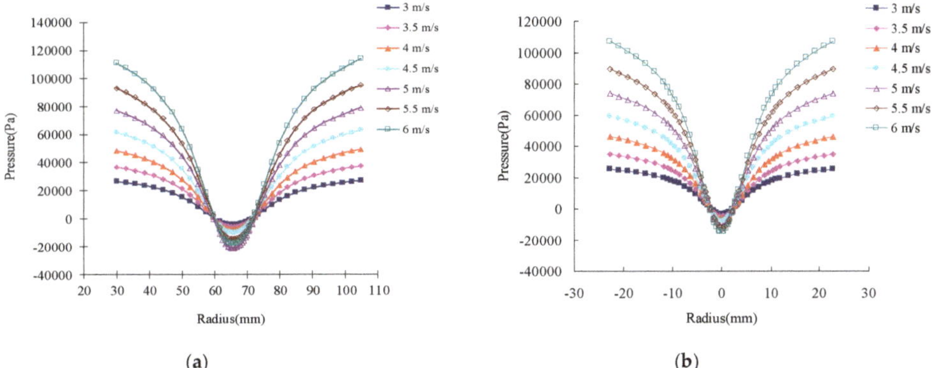

Figure 6. Radial pressure distributions: (**a**) $Z = 300$ mm (first stage); (**b**) $Z = 150$ mm (second stage).

The radial pressure gradient force $f_{\Delta p,i}$ is the main driving force that governs particle movement in the radial direction.

$$f_{\Delta p,i} = \frac{dp}{dr} V_{p,i} = \Delta p V_{p,i}, \tag{17}$$

where $V_{p,i}$ is the particle volume, Δp is the radial pressure gradient, and r is the radial distance from the center. Figure 6 shows that, as the inlet velocity increases, the absolute value of the negative pressure in the center of the hydrocyclone gradually increases, and the pressure near the walls also gradually increases. This is because the radial pressure in the semi-free vortex is proportional to the square of the tangential velocity at the position. The greater the inlet velocity of the hydrocyclone is, the faster the internal fluid flow velocity will be, and the higher the tangential velocity will be. Therefore, the radial pressure increases with the increase of the inlet velocity. In the forced vortex region, the centrifugal force field becomes stronger as the tangential velocity increases, such that lower negative pressure is generated in the center of the hydrocyclone. With the increase of tangential velocity, the strength of the centrifugal force field increases, and then the radial pressure gradient in the first stage of the hydrocyclone increases gradually with increasing inlet velocity. Equation (17) implies that the larger the radial pressure gradient is, the larger the force applied to the particles in the radial direction is, the easier it is for the particles to settle to the wall surface, and the easier it is for particles of the same size to enter the second stage of the hydrocyclone. The radial pressure gradient distribution law of the second stage of the hydrocyclone is the same as that of the first stage of the hydrocyclone. This is because, when the underflow of the first stage enters the second-stage hydrocyclone, it forms a similar rotational flow field. Meanwhile, as the inlet velocity increases, the velocity of the fluid entering the second stage increases accordingly. Therefore, the pressure distribution rule in the second-stage hydrocyclone is similar to that of the first-stage hydrocyclone. That is, the radial pressure gradient increases gradually with increasing inlet velocity.

3.2.2. Tangential Velocity

For the fluid flow in the hydrocyclone, the tangential velocity directly determines the strength of the centrifugal force field $f_{c,i}$, which is the main driving force for phase separation and is an important factor for evaluating separation performance.

$$f_{c,i} = \Delta V_{p,i} \rho_i \frac{u_t^2}{r}, \tag{18}$$

where ρ_i is the particle density. Equation (18) indicates that the centrifugal force is proportional to the tangential velocity. Analyses of tangential velocity were also performed at the $Z = 300$ mm cross-section and the $Z = 150$ mm cross-section, shown in Figure 7. The tangential velocity conforms to the radial symmetric Rankin vortex distribution. From the axial center to the walls, the tangential velocity firstly increases sharply along the radius, which is in accordance with the tangential velocity distribution of the forced vortex with a large velocity gradient. After reaching its maximum value, the tangential velocity begins decreasing with increasing radius and rapidly decreases to zero near the walls, which is in accordance with the tangential velocity distribution of the free vortex.

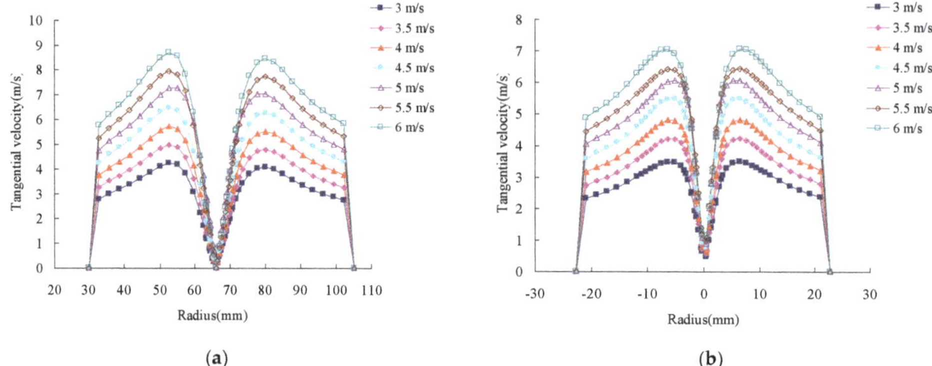

(a) (b)

Figure 7. Comparison of tangential velocities: (**a**) Z = 300 mm (first stage); (**b**) Z = 150 mm (second stage).

It can also be found that the inlet velocity has great influence on the tangential velocity. When the inlet velocity increases from 3 m/s to 6 m/s, the maximum tangential velocity of the first stage of the hydrocyclone increases from 4.23 m/s to 8.70 m/s, and the maximum tangential velocity of the second stage of the hydrocyclone increases from 3.49 m/s to 7.07 m/s. This can be attributed to the fact that, as a velocity component, the tangential velocity increases inevitably with the increasing inlet velocity. The maximum tangential velocity in the first stage of the hydrocyclone is approximately 1.43 times the inlet velocity, while the maximum tangential velocity in the second stage of the hydrocyclone is only approximately 0.83 times the first-stage inlet velocity.

Equation (18) shows that the centrifugal force on a particle is proportional to the tangential velocity at the particle's position. The greater the tangential velocity is, the stronger the centrifugal force field is, and the greater the centrifugal force applied to the particles is. The separation of the particles is mainly affected by the radial centrifugal force. Hence, the larger the tangential velocity is, the easier it is for the particles to overcome the fluid resistance to move toward the walls. In the first stage of the hydrocyclone, the particles are subjected to the strongest centrifugal force. Most of the particles can overcome the centrifugal force and enter the second stage of the hydrocyclone. Only a small subset of the finest particles are contained in the first-stage overflow. After entering the second stage of the hydrocyclone, the particles are separated again. Under the relatively weak centrifugal force, coarse particles directly enter the second-stage underflow, while the remaining middling particles are discharged from the second-stage overflow.

3.2.3. Axial Velocity

The fluid velocity in the hydrocyclone can be divided into axial velocity, tangential velocity, and radial velocity. The radial velocity reflects the motion of the fluid along the radial direction and the tangential velocity reflects the rotation of the fluid, while the axial velocity reflects the motion along the axial direction. The fluid inside the hydrocyclone enters the underflow from top to bottom or enters the overflow from bottom to top. If the structural dimensions of the hydrocyclone are unchanged, the axial velocity will determine the total time from entry to exit of the fluid. In other words, the axial velocity affects the residence time of the fluid and the split ratio, which in turn influences the hydrocyclone's separation performance.

Axial velocity can also reflect the direction of fluid flow in the hydrocyclone. Figure 8 shows the axial velocity at the Z = 300 mm cross-section and the Z = 150 mm cross-section; it can be found that, in the first stage of the hydrocyclone, the fluid is divided into three regions based on the axial velocities. Near the sidewall, the axial velocity is negative, and the fluid flows downward to the underflow. This region is referred to as the outer vortex region. Near the center of the hydrocyclone, the axial velocity is positive and the fluid flows upward to the overflow. This region is referred to as the inner vortex region. In the central region of the hydrocyclone, the axial velocity is negative. This is because the pressure in the central region is lower than the external atmospheric pressure. Outside air tends to flow into the hydrocyclone since the first-stage vortex finder is open to the atmosphere.

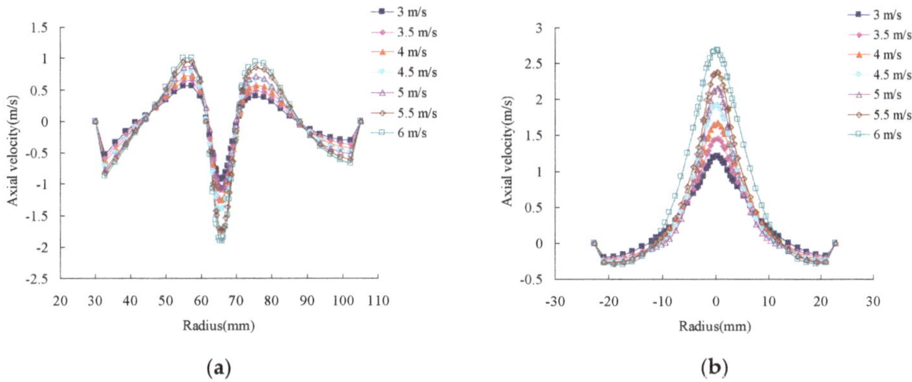

Figure 8. Axial velocity profiles: (**a**) Z = 300 mm (first stage); (**b**) Z = 150 mm (second stage).

Moreover, as the inlet velocity increases, the axial velocity in the central negative pressure region of the first stage of the hydrocyclone gradually increases, and the negative pressure region gradually extends downward. In the first stage of the hydrocyclone, the axial velocities of the outer vortex and inner vortex both increase as the inlet velocity increases. In the second stage of the hydrocyclone, when the inlet velocity increases, the axial velocity of the outer vortex gradually increases while the axial velocity of the inner vortex remains substantially unchanged. The axial velocity in the central region of the second stage of the hydrocyclone increases significantly with increasing inlet velocity.

3.2.4. Radial Velocity

Compared to the tangential and axial velocities, the radial velocity is relatively small, but its distribution is complex. During the separation process, the radial velocity has a relatively small effect on the centrifugal force field, but it can affect the radial movement of the particles. Figure 9 shows the radial velocity contours at the Z = 300 mm and Z = 150 mm cross-sections of the TS hydrocyclone. The radial velocity gradient increases as the inlet velocity increases, which is beneficial to the radial separation of the particles. The radial velocity in the first stage of the hydrocyclone is distributed with high symmetry. The velocities are all directed from the wall to the axial center. As the radius decreases, the radial velocity first increases gradually, and then decreases rapidly after reaching its maximum value near the axial center, which conforms to the radial velocity distribution of the sink flow. The second stage of the hydrocyclone has the conditions for forming the air core because of the decrease in velocity and the interconnection between the vortex finder and the apex, which results in the radial velocity's poor symmetry.

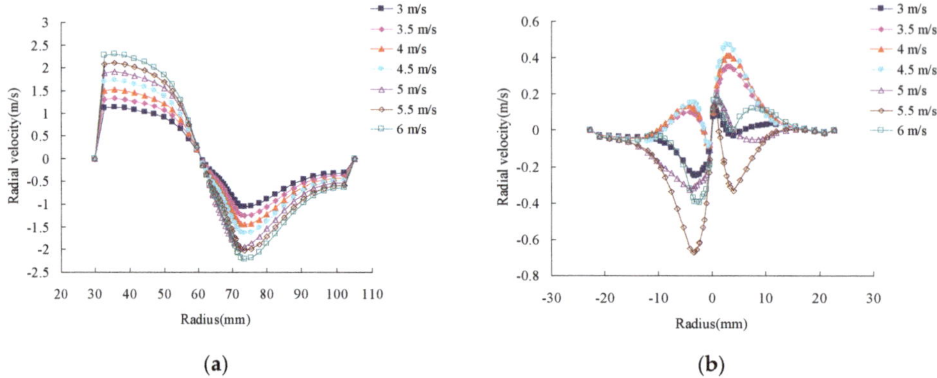

Figure 9. Radial velocity profiles: (**a**) $Z = 300$ mm (first stage); (**b**) $Z = 150$ mm (second stage).

3.3. Separation Performance

3.3.1. Classification Efficiency

The classification efficiency is defined as the ratio of particles with a certain size discharged through the underflow to the total particles with that size in the feed, i.e., the separation efficiency for a specific particle grade. The classification efficiency for particles of different sizes was firstly determined. Then, the grade efficiency curves were obtained by plotting the relationship curves between the particle size and the classification efficiency. The classification effect of the hydrocyclone can be directly reflected by the grade efficiency curves. Figure 10 shows the influence of the inlet velocity on the grade efficiency curves of the TS hydrocyclone. The grade efficiency curves all gradually shift to the left as the inlet velocity increases. This is because the first-stage tangential velocity increases and the centrifugal force field strengthens as the inlet velocity increases. Under the influence of the high-intensity centrifugal force field, more particles move toward the sidewall and enter the second stage of the hydrocyclone. The tangential velocity in the second stage of the hydrocyclone also increases with the increasing inlet velocity. The centrifugal force in the second stage of the hydrocyclone also becomes stronger, which causes particle recoveries to increase.

Figure 10 also shows that the classification efficiency of particles smaller than 10 μm is not affected by the inlet velocity within the scope of this study. This is because particles smaller than 10 μm are affected less by the centrifugal force field. It is difficult for these particles to overcome the fluid resistance. Therefore, the separation efficiency of fine particles that smaller than 10 μm does not change significantly with increases in inlet velocity. Further analysis shows that the first-stage hydrocyclone has little separation effect for fine particles, and only 50% of particles smaller than 10 μm can be separated to the underflow. However, it has obvious separation efficiency for coarse particles, and almost all the particles larger than 50 μm can be separated to the underflow. It can be concluded that the main function of the first-stage hydrocyclone is to separate most coarse particles to the second-stage hydrocyclone and guarantee that no coarse particles mixed in the overflow product of the first stage. From Figure 10b, the second-stage hydrocyclone has obvious separation effect for fine particles. Only 2% of particles smaller than 10 μm enter the underflow of the second-stage hydrocyclone. However, it cannot completely recover coarse particles larger than 50 μm. Therefore, the main function of the second-stage hydrocyclone is to separate most fine particles to the overflow and guarantee that no fine particles mixed in the underflow product.

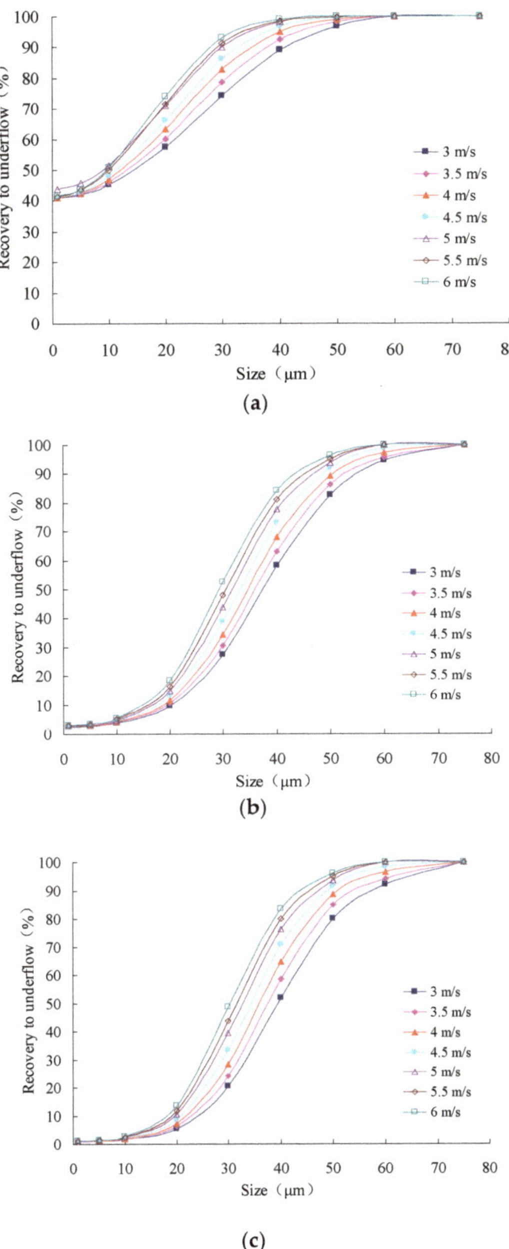

Figure 10. Separation efficiency: (**a**) first stage; (**b**) second stage; (**c**) TS hydrocyclone.

The separation performance of the TS hydrocyclone can be visually evaluated by analyzing the efficiency curves. The cut size d_{50}, the overall separation efficiency, the split ratio, and the possible deviation E_p of the TS hydrocyclone are shown in Figure 11. As the inlet velocity increases from 3 m/s to 6 m/s, the cut size d_{50} is reduced linearly from 39.5 μm to 30.5 μm, and the overall separation efficiencies of the first- and second-stage overflows are both reduced as the inlet velocity increases. This is because, as the velocity increases, the centrifugal force field becomes stronger and more particles can overcome the fluid resistance and enter the underflow. As shown in Figure 11b, the overall separation efficiency of the particles is reduced by 39.99% for the first-stage overflow and by 18.56% for the second-stage overflow. In contrast, the overall recovery efficiency of the second-stage underflow is increased by 56.54%. The results are consistent with those obtained from the efficiency curve analysis. Figure 11c presents the relationship between the inlet velocity and the liquid flow at each outlet of the TS hydrocyclone. As the inlet velocity increases, the first-stage and second-stage overflows are both reduced slightly, while the second-stage underflow is slightly increased; however, the magnitudes of the changes are all small. This is because, when the inlet velocity increases, the axial velocity of the external swirl of the first-stage cyclone and the second-stage cyclone both increase slightly. Therefore, the inflow into the underflow increases and the overflow flow decreases. However, since the axial velocity is less affected by the inlet velocity, the change is not obvious. Figure 11d reflects the change of the possible deviation E_p of the TS hydrocyclone after the inlet velocity is increased. Within the scope of this study, the larger the inlet velocity is, the smaller the possible deviation E_p is, and the higher the separation sharpness is.

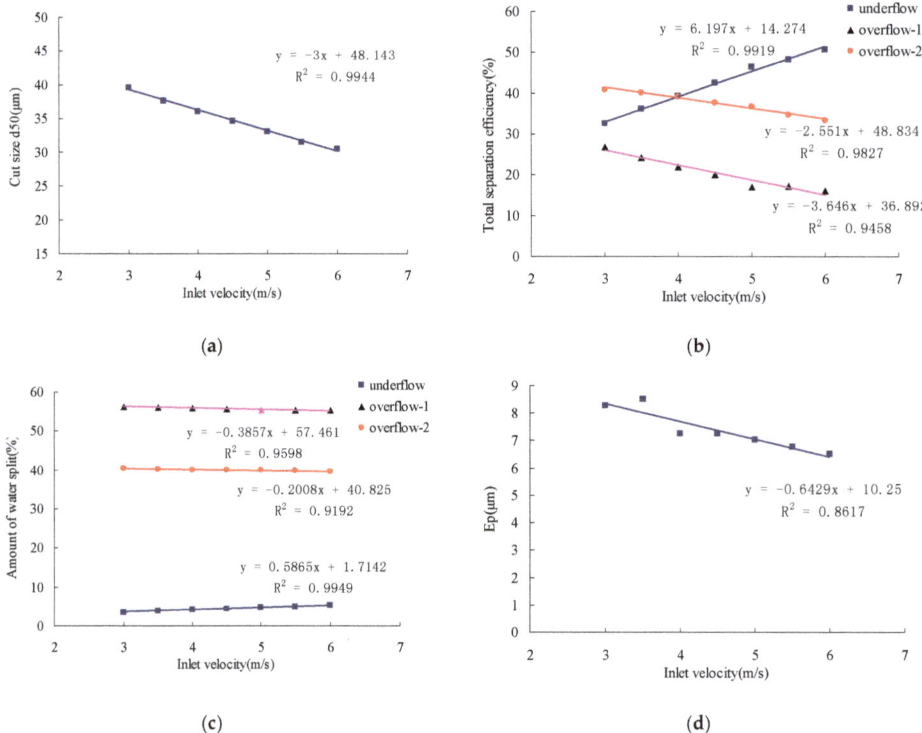

Figure 11. Comparison of separation performance: (**a**) cut size d_{50}; (**b**) total separation efficiency; (**c**) split ratio; (**d**) E_p.

3.3.2. Particle Size

The particle size distribution of the classified product is a direct indicator for evaluating the hydrocyclone's separation performance. Therefore, the particle size of each product was analyzed, and the results are shown in Figure 12. The particle size of the first-stage overflow product is significantly different from that of the feed. Particles larger than 40 μm in the feed are all discharged to the second stage. The first-stage overflow is substantially free of coarse particles, which indicates an obvious classification effect. After the two-stage hydrocyclone separation, the second-stage underflow is free of particles smaller than 10 μm and almost free of particles smaller than 20 μm. The second-stage underflow contains few fine particles. The second-stage overflow product consists of middling particles, but still includes some fine particles and a small number of coarse particles. The particle sizes in the second-stage overflow product are similar to those in the feed.

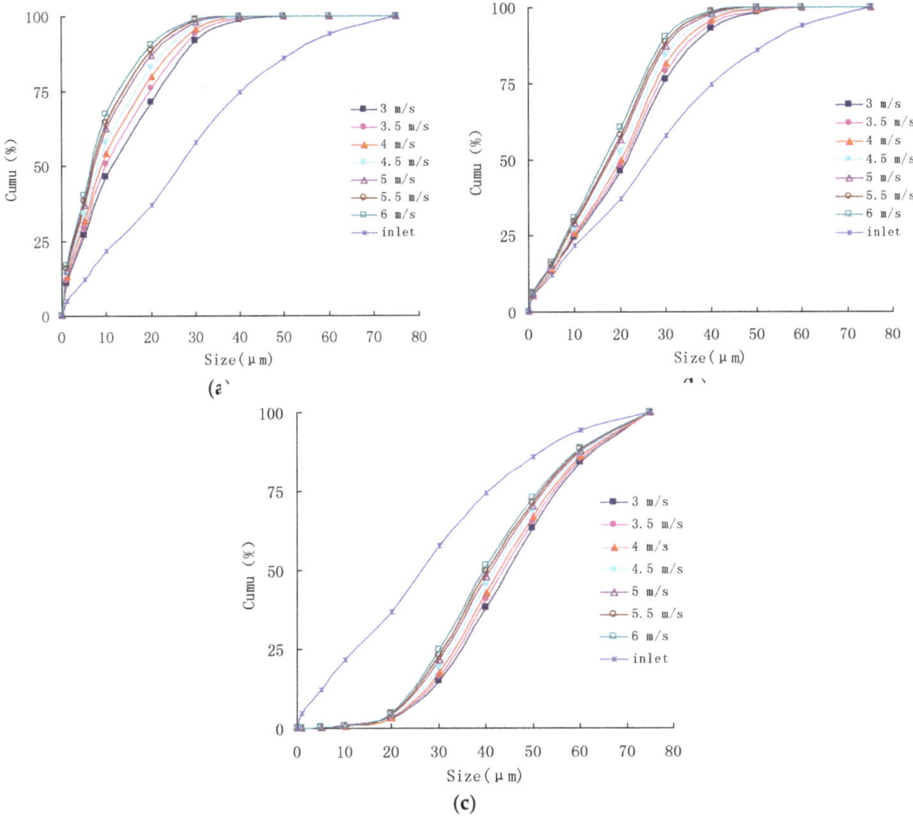

Figure 12. Particle size distribution of the product: (**a**) the overflow of first stage; (**b**) the overflow of second stage; (**c**) the underflow of second stage.

As the inlet velocity increases, the particle size of each product becomes smaller gradually. The reason is that the centrifugal force field is strengthened with the increasing tangential velocity of the fluid inside both stage hydrocyclones. As a result, some 20–40-μm particles which stay in the overflow of the first stage originally can overcome resistance and enter the second-stage hydrocyclone. Therefore, the particle size of the overflow product of the first stage becomes smaller due to loss of these particles. Then, these particles enter the second-stage hydrocyclone and discharge into the underflow. Compared to most particles inside the underflow product, these particles can be regarded as fine particles; therefore, the overall particle size of the underflow also becomes smaller.

In addition to the particle size distribution, a representative particle size can also be used to characterize the particle size in each product. One of the commonly used indicators is the median particle size, which is the particle size at the point corresponding to 50% on the cumulative particle size distribution curve of each product. The median particle size of each product is shown in Figure 13. As the inlet velocity increases from 3 m/s to 6 m/s, the median particle size in the first-stage overflow decreases from 11 μm to 6.5 μm, with a reduction of 40.91%. The median particle size in the second-stage overflow decreases from 21.5 μm to 16.5 μm, with a reduction of 23.26%. The median particle size in the second-stage overflow decreases from 44.75 μm to 39.5 μm, with a reduction of 11.73%. Moreover, the median particle size of the second-stage overflow product is close to that of the feed (26 μm). Inspired by this observation, we propose recycling the second-stage overflow to the feed for re-classification and using only the first-stage overflow and the second-stage underflow as products. In this way, products free of particle entrainments can be obtained, and the goal of precise classification can be achieved.

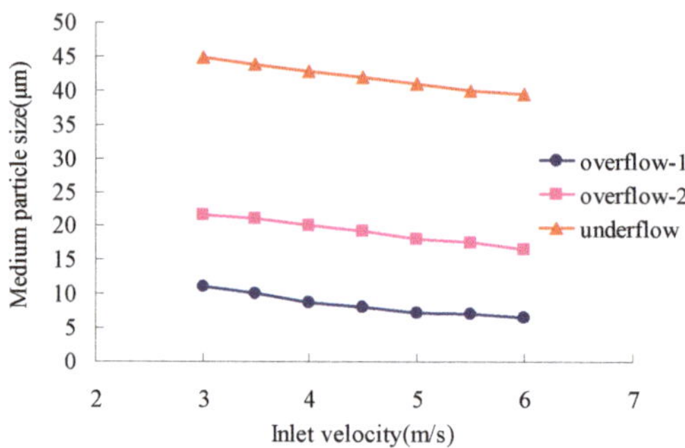

Figure 13. Variation of the medium particle size of the product.

4. Conclusions

In this study, a two-stage series double efficiency hydrocyclone was designed. The effects of inlet velocity on the internal flow field and on the separation characteristics of the TS hydrocyclone were investigated by CFD. The major conclusions are summarized as follows:

(1) The pressure drop of the TS hydrocyclone increases linearly as the inlet velocity increases. When the inlet velocity increases from 3 m/s to 6 m/s, the pressure drop of the first stage of the hydrocyclone increases from 27.81 kPa to 116.21 kPa, and the pressure drop of the second stage of the hydrocyclone increases from 26.14 kPa to 111.17 kPa.

(2) When the inlet velocity increases from 3 m/s to 6 m/s, the maximum tangential velocity of the first stage of the hydrocyclone increases from 4.23 m/s to 8.7 m/s, and the maximum tangential velocity of the second stage of the hydrocyclone increases from 3.49 m/s to 7.07 m/s. The maximum tangential velocity in the first stage of the hydrocyclone is approximately 1.43 times the inlet velocity, while the maximum tangential velocity in the second stage of the hydrocyclone is only approximately 0.83 times the first-stage inlet velocity.

(3) The overall separation efficiencies of the first- and second-stage overflows both decrease as the inlet velocity increases. The overall separation efficiency of the particles is reduced by 39.99% for the first-stage overflow and by 18.56% for the second-stage overflow. In contrast, the overall recovery efficiency of the second-stage underflow increases by 56.54%.

(4) In the scope of this study, the larger the inlet velocity is, the smaller the cut size d_{50} is, the smaller the possible deviation E_p is, and the higher the separation sharpness of the TS hydrocyclone is.

(5) After the two-stage separation, the first-stage overflow contains few coarse particles above 40 μm and the second-stage underflow contains few fine particles. The second-stage underflow is free of particles smaller than 10 μm and almost free of particles smaller than 20 μm. The underflow product contains few fine particles.

Author Contributions: Conceptualization, P.L.; data curation, Y.Z. and H.W.; formal analysis, Y.Z.; investigation, X.Y.; methodology, L.J.; resources, P.L.; software, L.J.; validation, X.Y.; visualization, H.W.; writing—original draft, L.J.; writing—review and editing, L.J., P.L., and X.Y.

Funding: This work was supported by the National Key R&D Program of China (2018YFC0604702), and the Shandong Provincial Key Research and Development Program, China (2017GSF216004).

Conflicts of Interest: The authors declare no conflicts of interest.

References

1. Ghodrat, M.; Qi, Z.; Kuang, S.B.; Ji, L.; Yu, A.B. Computational investigation of the effect of particle density on the multiphase flows and performance of hydrocyclone. *Miner. Eng.* **2016**, *90*, 55–69. [CrossRef]
2. Hwang, K.J.; Hwang, Y.W.; Yoshida, H.; Shigemori, K. Improvement of particle separation efficiency by installing conical top-plate in hydrocyclone. *Powder Technol.* **2012**, *232*, 41–48. [CrossRef]
3. Delgadillo, J.A. Modelling of 75- and 250-mm Hydrocyclones and Exploration of Novel Designs Using Computational Fluid Dynamics. Ph.D. Thesis, Department of Metallurgical Engineering, University of Utah, Salt Lake City, UT, USA, 2006.
4. Slack, M.D.; Porte, S.D.; Engelman, M.S. Designing automated computational fluid dynamics modelling tools for hydrocyclone design. *Miner. Eng.* **2004**, *17*, 705–711. [CrossRef]
5. Olson, T.J.; Ommen, R.V. Optimizing hydrocyclone design using advanced CFD model. *Miner. Eng.* **2004**, *17*, 713–720. [CrossRef]
6. Motin, A.; Tarabara, V.V.; Petty, C.A.; Bénard, A. Hydrodynamics within flooded hydrocyclones during excursion in the feed rate: Understanding of turndown ratio. *Sep. Purif. Technol.* **2017**, *185*, 41–53. [CrossRef]
7. Ni, L.; Tian, J.Y.; Zhao, J.N. Experimental study of the effect of underflow pipe diameter on separation performance of a novel de-foulant hydrocyclone with continuous underflow and reflux function. *Sep. Purif. Technol.* **2016**, *171*, 270–279. [CrossRef]
8. Wang, J.G.; Bai, Z.S.; Yang, Q.; Fan, Y.; Wang, H.L. Investigation of the simultaneous volumetric 3-component flow field inside a hydrocyclone. *Sep. Purif. Technol.* **2016**, *163*, 120–127. [CrossRef]
9. Ji, L.; Kuang, S.; Qi, Z.; Wang, Y.; Chen, J.; Yu, A.B. Computational analysis and optimization of hydrocyclone size to mitigate adverse effect of particle density. *Sep. Purif. Technol.* **2017**, *174*, 251–263. [CrossRef]
10. Hwang, K.J.; Chou, S.P. Designing vortex finder structure for improving the particle separation efficiency of a hydrocyclone. *Sep. Purif. Technol.* **2017**, *172*, 76–84. [CrossRef]
11. Cui, B.Y.; Wei, D.Z.; Gao, S.L.; Liu, W.G.; Feng, Y.Q. Numerical and experimental studies of flow field in hydrocyclone with air core. *Trans. Nonferrous Met. Soc. China* **2014**, *24*, 2642–2649. [CrossRef]
12. Wang, B.; Yu, A.B. Numerical study of the gas-liquid-solid flow in hydrocyclones with different configuration of vortex finder. *Chem. Eng. J.* **2008**, *135*, 33–42. [CrossRef]
13. Jiang, L.Y.; Liu, P.K.; Zhang, Y.K.; Yang, X.H.; Wang, H.; Gui, X.H. Design boundary layer structure for improving the particle separation performance of a hydrocyclone. *Powder Technol.* **2019**, *350*, 1–14. [CrossRef]
14. Pei, B.B.; Yang, L.; Dong, K.J.; Jiang, Y.C.; Du, X.S.; Wang, B. The effect of cross-shaped vortex finder on the performance of cyclone separator. *Powder Technol.* **2017**, *313*, 135–144. [CrossRef]
15. Ghodrat, M.; Kuang, S.B.; Yu, A.B.; Vince, A.; Barnett, G.D.; Barnett, P.J. Numerical analysis of hydrocyclones with different conical section designs. *Miner. Eng.* **2014**, *62*, 74–84. [CrossRef]
16. Vieira, L.G.; Damasceno, J.J.; Barrozo, M.A. Improvement of hydrocyclone separation performance by incorporating a conical filtering wall. *Chem. Eng. Process.* **2010**, *49*, 460–467. [CrossRef]
17. Silva, N.K.; Silva, D.O.; Vieira, L.G.; Barrozo, M.A. Effects of underflow diameter and vortex finder length on the performance of a newly designed filtering hydrocyclone. *Powder Technol.* **2015**, *286*, 305–310. [CrossRef]
18. Restarick, C.J. Classification with two-stage cylinder-cyclones in small-scale grinding and flotation circuits. *Int. J. Miner. Process.* **1989**, *26*, 165–179. [CrossRef]
19. Restarick, C.J. Adjustable onstream classification using a two-stage cylinder-cyclone. *Miner. Eng.* **1991**, *4*, 279–288. [CrossRef]
20. Zhang, X.M.; Guo, D. Determination of design parameter for three-product heavy-medium cyclone. *J. Coal Sci. Eng.* **2011**, *17*, 96–99. [CrossRef]
21. Obeng, D.P.; Morrell, S. The JK three-product cyclone-performance and potential applications. *Int. J. Miner. Process.* **2003**, *69*, 129–142. [CrossRef]
22. Obeng, D.P.; Morrell, S.; Napier-Munn, T.J. Application of central composite rotatable design to modelling the effect of some operating variables on the performance of the three-product cyclone. *Int. J. Miner. Process.* **2005**, *76*, 181–192. [CrossRef]

23. Ahmed, M.M.; Ibrahim, G.A.; Farghaly, M.G. Performance of a three-product hydrocyclone. *Int. J. Miner. Process.* **2009**, *91*, 34–40. [CrossRef]
24. Krokhina, A.V.; Dueck, J.; Neesse, T.; Min'kov, L.L.; Pavlikhin, G.I. An Investigation into Hydrodynamics of a Hydrocyclone with an Additional Double Jet Injector. *Theor. Found. Chem. Eng.* **2011**, *45*, 213–220. [CrossRef]
25. Farghaly, M.G.; Golyk, V.; Ibrahim, G.A.; Ahmed, M.M.; Neesse, T. Controlled wash water injection to the hydrocyclone underflow. *Miner. Eng.* **2010**, *23*, 321–325. [CrossRef]
26. Golyk, V.; Huber, S.; Farghaly, M.G.; Prölss, G.; Endres, E.; Neesse, T.; Hararah, M.A. Higher kaolin recovery with a water-injection cyclone. *Miner. Eng.* **2011**, *24*, 98–101. [CrossRef]
27. Dueck, J.; Pikushchak, E.; Minkov, L.; Farghaly, M.; Neesse, T. Mechanism of hydrocyclone separation with water injection. *Miner. Eng.* **2010**, *23*, 289–294. [CrossRef]
28. Wang, C.C.; Wu, R.M. Experimental and simulation of a novel hydrocyclone-tubular membrane as overflow pipe. *Sep. Purif. Technol.* **2018**, *198*, 60–67. [CrossRef]
29. Garcia, D.V.; Parada, P.R.B.; Cilliers, J.J. Optimising small hydrocyclone design using 3D printing and CFD simulations. *Chem. Eng. J.* **2018**, *350*, 653–659. [CrossRef]
30. Chiné, B.; Concha, F. Flow patterns in conical and cylindrical hydrocyclones. *Chem. Eng. J.* **2000**, *80*, 267–273. [CrossRef]
31. Delgadillo, J.A.; Rajamani, R.K. Exploration of hydrocyclone designs using computational fluid dynamics. *Int. J. Miner. Process.* **2007**, *84*, 252–261. [CrossRef]
32. Narasimha, M.; Brennan, M.S.; Holtham, P.N. CFD modeling of hydrocyclones: Prediction of particle size segregation. *Miner. Eng.* **2012**, *39*, 173–183. [CrossRef]
33. Brennan, M. CFD simulations of hydrocyclones with an air core: Comparison between large eddy simulations and a second moment closure. *Chem. Eng. Res. Des.* **2006**, *84*, 495–505. [CrossRef]
34. Slack, M.D.; Prasad, R.O.; Bakker, A.; Boysan, F. Advances in cyclone modeling using unstructured grids. *Chem. Eng. Res. Des.* **2000**, *78*, 1098–1104. [CrossRef]
35. Vakamalla, T.R.; Mangadoddy, N. Numerical simulation of industrial hydrocyclones performance: Role of turbulence modelling. *Sep. Purif. Technol.* **2017**, *176*, 23–39. [CrossRef]
36. Chu, K.W.; Wang, B.; Yu, A.B.; Vince, A. CFD-DEM modelling of multiphase flow in dense medium cyclones. *Powder Technol.* **2009**, *193*, 235–247. [CrossRef]
37. Kyriakidis, Y.N.; Silva, D.O.; Barrozo, M.A.; Vieira, L.G. Effect of variables related to the separation performance of a hydrocyclone with unprecedented geometric relationships. *Powder Technol.* **2018**, *338*, 645–653. [CrossRef]
38. Hsieh, K.T.; Rajamani, K. Phenomenological model of hydrocyclone: Model development and verification for single-phase flow. *Int. J. Miner. Process.* **1988**, *22*, 223–237. [CrossRef]

Article

Application of the Discrete Element Method to Study the Effects of Stream Characteristics on Screening Performance

Ali Davoodi *, Gauti Asbjörnsson, Erik Hulthén and Magnus Evertsson

Department of Industrial and Materials Science, Chalmers University of Technology, 412 96 Gothenburg, Sweden; gauti@chalmers.se (G.A.); erik.hulthen@chalmers.se (E.H.); magnus.evertsson@chalmers.se (M.E.)
* Correspondence: ali.davoodi@chalmers.se

Received: 24 October 2019; Accepted: 11 December 2019; Published: 14 December 2019

Abstract: Screening is a key operation in a crushing plant that ensures adequate product quality of aggregates in mineral processing. The screening process can be divided into the two sub-processes of stratification and passage. The stratification process is affected by the relative difference between various properties, such as particle shape, size distribution, and material density. The discrete element method (DEM) is a suitable method for analyzing the interactions between individual particles and between particles and a screen deck in a controlled environment. The main benefit of using the DEM for simulating the screening process is that this method enables the tracking of individual particles in the material flow, and all of the collisions between particles and between particles and boundaries. This paper presents how different particle densities and flowrates affect material stratification and, in turn, the screening performance. The results of this study show that higher density particles have a higher probability of passage because of their higher stratification rate, which increases the probability that a particle will contact the screen deck during the process.

Keywords: screen; DEM; discrete element method; classification efficiency; separation

---

1. Introduction

Screening is a key unit operation in a crushing plant that ensures adequate product quality of aggregates in mineral processing. The main purpose of screening is to classify particles based on their shape and size. Efficient screening can reduce operational costs by improving product structure, product quality, and energy efficiency [1].

During the screening process, many operational parameters can affect the performance of screening. These parameters can be divided into either machine parameters or stream characteristics. Machine parameters, such as the screen dimensions, deck material, vibration frequency, vibration amplitude, and inclination, are dependent on the installed unit and selected operational strategy. The stream characteristics include material properties such as the size distribution, shape, density, and flow rate of the material [2].

During operation, a material bed builds upon each deck. Particles that are smaller than the aperture and are in contact with the deck have a certain probability of passing through the aperture. Soldinger [3] proposed a mechanistic modeling approach to screening in which the granular flow was discretized along the deck and in horizontal layers to approximate horizontal and vertical transport. In the stratification process, the finer fractions change vertical placement with the larger fractions that are located in lower layers over a predefined time interval [4]. The passage rate is a function of the particle size distribution in the bottom layer of the bed, i.e., the particles that are in contact with the screen deck [4]. However, the model is not capable of approximating the change in the stratification rate as a function of the particle shape or density.

Granular materials with different properties exhibit different segregation behaviors, whether in a vibrating or flowing bed. The three most important material properties that can affect segregation are particle shape, size, and density [5].

Different models have been proposed for predicting segregation in granular flows. In a research study by Khakhar et al. [6], a model for the segregation flux in flowing layers was presented and validated by particle dynamics and Monte Carlo simulations for steady flow down an inclined plane. Bridgwater et al. [7] developed a theory and used a mathematical model to describe the segregation of granular material; the authors also developed a model with a segregation velocity based on separation due to particle size differences, and in the model proposed by Wang et al. [8], the particle movement was considered. In work by Clément et al. [9], experiments on the motion of single particles of different sizes in a bed were done by using an image-processing device. In a study by Xiao et al. [10], both the discrete element method (DEM) and experiments were used to model density segregation in a flowing material. Only a few works have been completed by using continuum methodologies. In work by Carvalho and Tavares [11], an efficient method was presented for solving the two-way coupling of DEM and a mechanistic breakage model, to describe continuous grinding in a pilot-scale mil.

The granular flow of particles in the screening process is dependent on the interactions between individual particles, and between particles and the screen deck, which can be difficult to quantify experimentally. Simulation has become a common tool in the design and optimization of the processing of granular materials [12]. Of the simulation methods that have been used recently, one is the discrete element method (DEM). DEM is a simulation method that is discontinuous, which is required in granular material simulations because of the highly discontinuous nature of these materials [13]. Several studies have been conducted using DEM simulations to analyze separation performance by studying different parameters and particle behavior during screening operation. For example, the effect of particle size distribution was studied by Jahani et al. [14], and the effects of the aperture shapes and materials of screen decks on the screening efficiency were studied by Davoodi et al. [15]. Additionally, the effect of different levels of acceleration was studied by et al. [16].

The objective of this paper is to study how different stream characteristics influence screening performance, in particular, the effect of the material density on stratification. In this paper, a model for the effect of the density on the stratification rate and performance is presented, and the approach is validated by comparing calibrated DEM simulations with an empirical model for vibratory screen. Table 1 shows the description of the parameters which have been used in this paper.

Table 1. Nomenclature.

F_n	Normal force, Equation (1), (N)	M_{i,j,n_l}	The mass flow along the screen, Equation (4), (kg)
K_n	Stiffness of the spring, Equation (1), (-)	M_{down,i,j,n_l-1}	The downward flow, Equation (4), (kg)
C_n	Viscoelastic damping constant, Equation (1), (m/s)	M_{up,i,j,n_l}	The upward flow, Equation (4), (kg)
V_n	The relative velocities, Equation (1), (m/s)	$M_{BP,i,j}$	Mass flow of the material in the contact layer, Equation (4), (kg)
F_t	Tangential force, Equation (2), (N)	k_j	Passage rate parameter, Equation (4), (-)
K_t	Stiffness of the spring, Equation (2), (-)	Δt	Time step, Equation (4), (s)
C_t	Viscoelastic damping constant, Equation (2), (-)	α	Slope of the screen deck, Equation (6), (Angle)
V_t	The relative velocities, Equation (2), (m/s)	v	Transport velocity of the particle along the screen deck, Equation (6), (m/s)
f	Frequency, Equations (3) and (6), (Hz)	R	Function of stroke, Equation (6), (mm)

2. Methodology

2.1. DEM

2.1.1. DEM Contact Model Theory

The DEM method uses a clear numerical structure to trace the motion of individual particles according to their interactions with each other and the system geometry (such as the screen deck and feeder), where a contact law is used to predict the velocities and rotations of the particles. There are various different contact models, such as the linear elastic and the Hertz–Mindlin contact models. However, the linear contact model may not be sufficiently precise for spherical particles. The Hertz–Mindlin contact model was developed to solve the contact behavior of spherical shapes. The Hertz–Mindlin contact model is a no-slip model with a linear spring-dashpot [17]. Figure 1 shows the interaction between two particles with frictional elements between the normal force and the tangential force [18]. The contact force between the impacting particles is split into a normal force F_n and a tangential force F_t as follows

$$F_n = -K_n \Delta x + C_n V_n \tag{1}$$

$$F_t = -K_t \Delta x + C_t V_t \tag{2}$$

where Δx denotes the particle displacements in the normal and tangential directions and V_n and V_t denote the relative velocities. K denotes the stiffness of the spring, and C denotes the viscoelastic damping constant. If F_t exceeds the limiting frictional force, then the particles will slide over each other, and the tangential force is calculated using the frictional coefficient f:

$$F_t = -fF_n \tag{3}$$

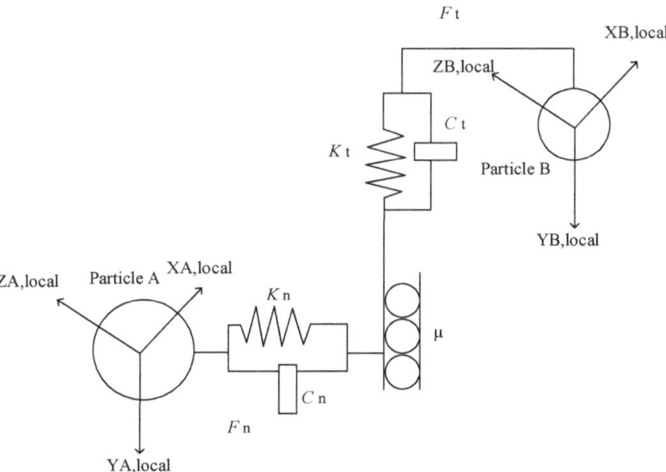

Figure 1. Graphic illustration of the Hertz–Mindlin contact model.

2.1.2. Simulations

The test plan was designed to determine how different material densities and altering the feed rate affect the screening efficiency. A total of nine simulations were run: three simulations for each different material density. As Table 2 shows, the feed rate was the parameter that was varied in the simulation and had initial values of 4 kg/s, 5 kg/s, and 6 kg/s. The particle size distribution (PSD), which, when generated for simulations are user-defined, meaning each of the particles are generated at different masses, as defined by the user. Figure 2 shows the PSD curve for the feed material in all simulations.

Table 2. of simulation parameters.

Material Properties	Poisson's Ratio	Shear Modulus	Density
Particles (Low density)	0.3	24 MPa	2100 kg/m^3
Particles (Medium density)	0.3	24 MPa	2500 kg/m^3
Particles (High density)	0.3	24 MPa	2900 kg/m^3
Screen (Steel)	0.2	79 GPa	7800 kg/m^3
Collision Properties	**Coefficient of Restitution**	**Coefficient of Static Friction**	**Coefficient of Rolling Friction**
Particle-particle	0.2	0.6	0.01
Particle-screen (Steel)	0.6	0.45	0.01
Machine Parameters			
Screen aperture	25 mm×25 mm and 10 mm×10 mm		
Screen declination	15°		
Screen vibration	Sinusoidal translation, amplitude 6 mm, and frequency 13 Hz		
Particle generation rate	For particles at 4 kg/s, 5 kg/s, and 6 kg/s		

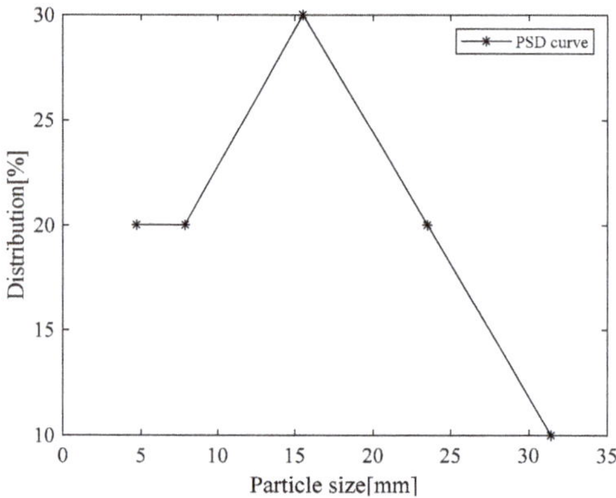

Figure 2. Particle Size Distribution (PSD) for feeding material.

The screen geometry that was used in simulations is based on a vibratory screen lab, a reduced-scale version of a real industry screen. The screen lab was used to minimize computing time by decreasing the geometrical area and the numbers of particles.

The next step was to set up the simulation. Particles with user-defined diameters were employed to match the planned simulations. All of the particles were non-spherical in shape, as very different results are obtained using one spherical particle in the simulations versus multi-sphere particles. Multi-sphere particles produce more realistic results because these particle shapes are closer to those of real particles [19]. Figure 3 shows the particle shape that was used in the simulations where all spheres had the same material properties. The material is a rock with three different solid densities of 2900 kg/m^3, 2500 kg/m^3, and 2100 kg/m^3. The separation of these particles may represent the essential sorting operation that occurs in a typical screening process. More generally, this model features the main characteristics of many industrial screening applications. Table 2 is a summary of the simulation parameters. The collision properties were calibrated by the method proposed by Quist [19].

Figure 3. Three spheres particle used in simulations.

Each DEM simulation should achieve a steady-state before data extraction; that is, the total number of particles in the system should become stable after a given period. The average time for the overflow particles to travel along the entire deck, from the feed end to the discharge end, was 5 s in the simulations, and the simulations reached a steady-state after 6 s, as can be seen in Figure 4.

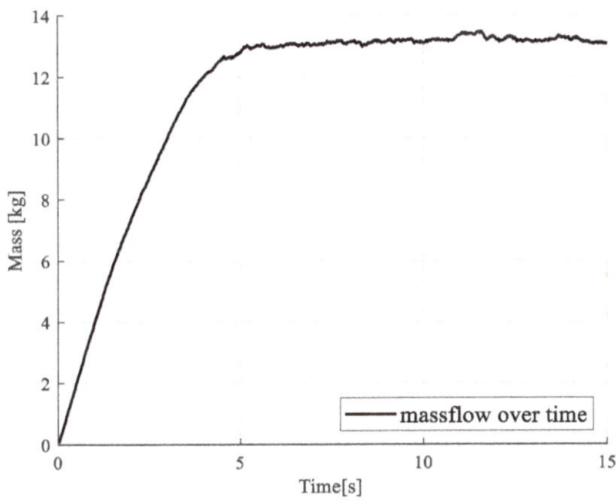

Figure 4. The mass flow over time in the simulation.

2.2. Laboratory Vibrating Screen

A laboratory-scale vibratory screen was constructed to control the screen motion, feed rate, and aperture size, and to enable the analysis of the particle size distribution. A schematic of the CAD geometry used in the simulations is shown in Figure 5. The screen was 1500 mm long and 300 mm wide.

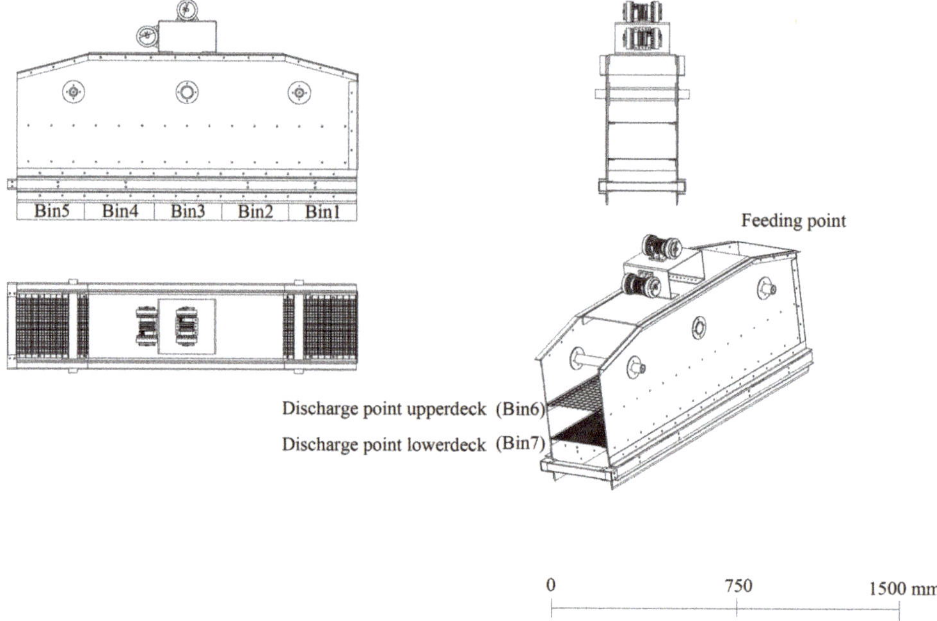

Figure 5. Schematic showing the CAD model used in the simulations.

2.3. Modeling

The screen model used in this study was based on the model proposed by Soldinger [20]. The material flow (M) was modeled as finite zones (i). Between the discrete zones, mass is transferred at each time step (Δt) in different fractions (j) in a layer (k). The mass flow is governed by Equation (4):

$$M_{i+1,j,n_l} = M_{i,j,n_l} + M_{down,i,j,n_l-1} - M_{up,i,j,n_l} - M_{BP,i,j}k_j\Delta t \tag{4}$$

Every zone (i) was divided into two layers, an upper and a lower layer. Between these layers, the material was exchanged through stratification. Figure 6 shows an example of a layer model for stratification and passage process for one screen deck.

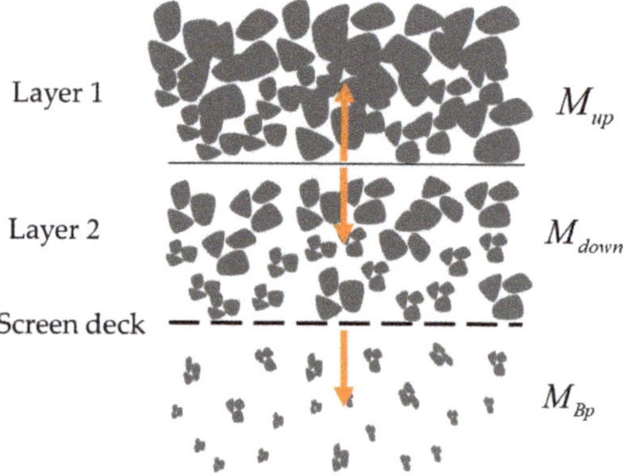

Figure 6. Layer model for stratification and passage process for one screen deck.

From the lower of these layers, material can pass through the screen apertures, which is determined by a passage probability (k), as described by Equation (5); which is determined by the average particle size in each fraction (d_{50}), the aperture size (Ap) and the rate factor (β). The rate factor can be calibrated to specific scenarios to provide more accurate results. The mass passing the aperture would be added to the zone below or added to the total mass passing all decks if there are no decks below the current one.

$$k_j = 80(e^{(-\beta d_{50}/Ap)} - e^{(-\beta)}d_{50}/Ap) \tag{5}$$

Material velocity can be determined by the inclination, frequency, and the throw of the screen. The input to the model was the angle (α) for every section. In the model, the incline of the screen changes the transport velocity of the rock material. Since a higher velocity would make the layers thinner, it would also change the stratification and passage through the screen decks. The velocity of the material was estimated with Equation (6). For a more detailed model description, see Soldinger [3] and Asbjörnsson et al. [2]. Figure 7 shows the mass balance between the different layers of vibratory.

$$v = (0.064\alpha + 0.2)(380R - 0.18)(0.095 f\alpha^{-0.5} + 0.018\alpha - 0.38) \tag{6}$$

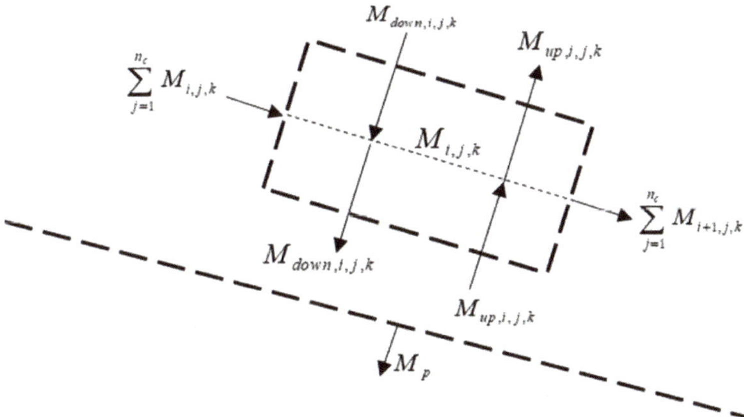

Figure 7. The mass balance between different layers of vibratory.

For the implementation of the model, a geometric design of the screen was configured, the feeding position was adjusted, the feed rate was fixed at 5 kg/s, and the size distribution was defined to match the DEM model.

3. Results/Discussion

The DEM simulation results for the vibratory screening process are shown in Figure 8, for which the feed rate increased from 4 kg/s to 6 kg/s. As the feed rate increased, a thicker, and approximately constant thickness material layer built upon the screen deck.

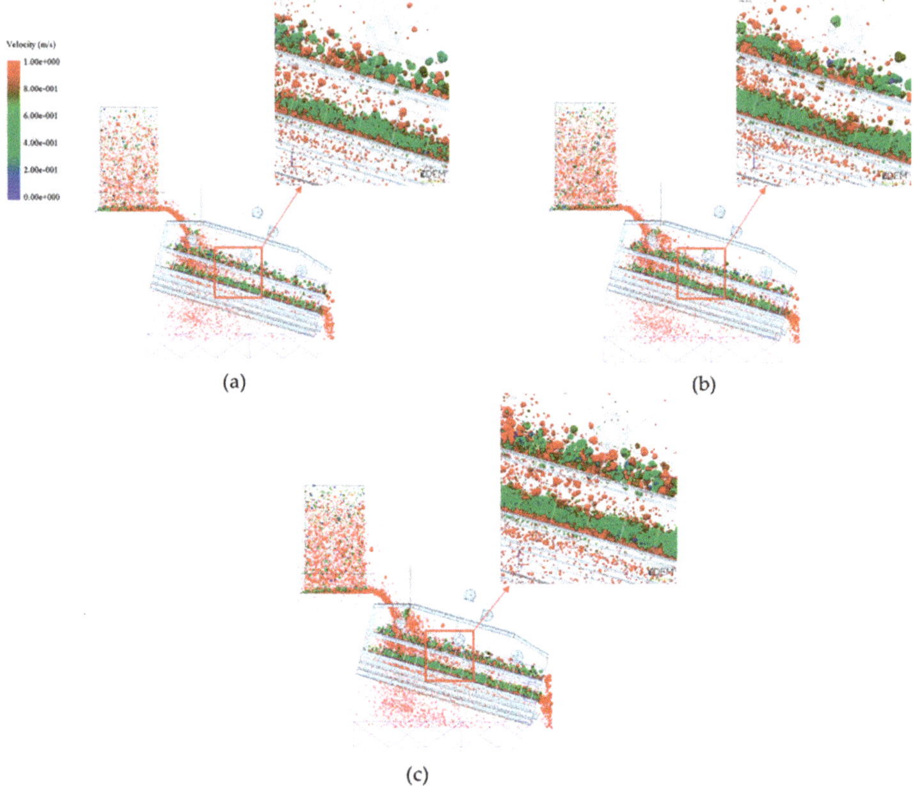

Figure 8. Discrete element method (DEM) simulation results for the vibratory screening process with three different feed rates: (**a**) 4 kg/s, (**b**) 5 kg/s, and (**c**) 6 kg/s.

The stratification process is different for various bed thicknesses. The stratification process that occurs during the screening process is expected to proceed in the vertical direction; thus, a thicker material bed requires more time for stratification. One way to adjust the stratification time is to control the particle velocity during the screening process by adjusting the inclination and frequency of the vibratory screen.

Stratification has the largest impact on screening efficiency. Stratification can be affected by different factors, such as material segregation, which itself depends on the physical particle characteristics, such as the particle size, shape, density, and surface texture. Particle size has a significant effect on segregation and stratification. Smaller particles can move towards the screen through the gaps between large particles, which is a key factor in screening efficiency.

In this paper, different sections of the vibratory screen were analyzed by using bin numbers to separate the sections. Material densities are simulated using three different feed rates. Figure 9 shows the percent of the particles (partition number) that passed through the different sections. The low-density material had a better passage than the high-density material.

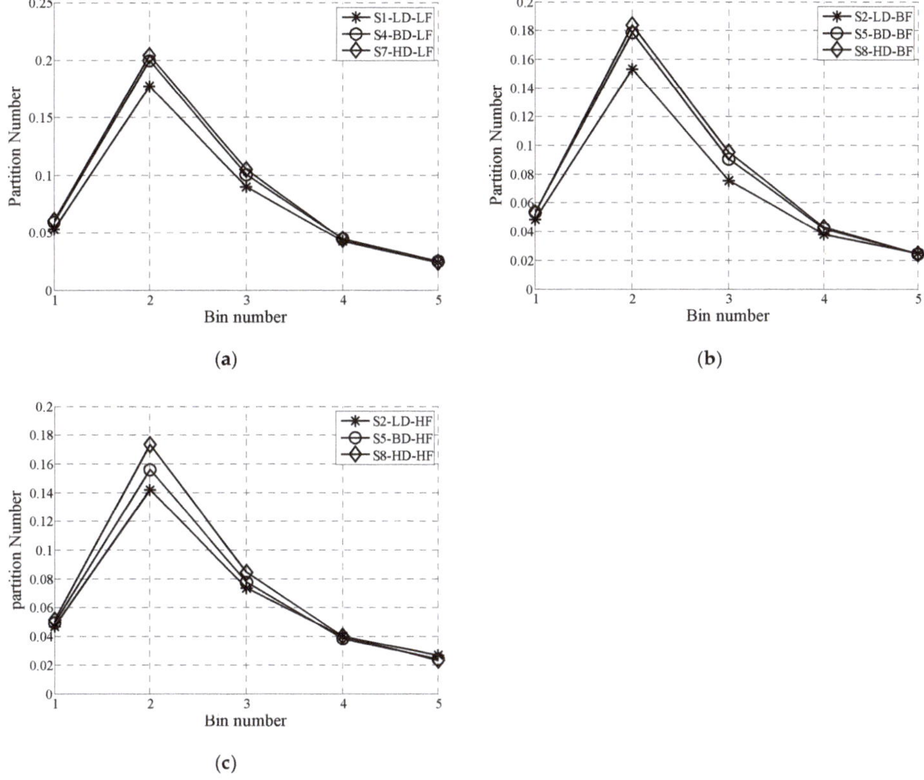

Figure 9. Passage rate in different screen sections for different densities and feed rates (LD: Low-density material, BD: Between density material, and HD: High-density material. (**a**) Low feed rate, 4 kg/s, (**b**) between feed rate, 5 kg/s, and (**c**) high feed rate, 6 kg/s.

For further investigation, the material discharge was studied to compare the numbers of undersized material in the overflow for both decks. The average particle diameter in the overflow material can be seen in Figure 10. The average diameter was lower for the low-density material for both decks, which means that the simulation of the low-density material had more undersized material in the overflow than high-density material. However, this difference was smaller on the upper deck than under the deck, since the upper deck had a larger aperture, and there were more chances of smaller particles passing through the screen deck.

Upon increasing the feed rate, the average particle diameters decreased for both decks (Figure 10), as increasing the number of particles in the simulation resulted in a thicker bed (Figure 8) and decreased the probability that smaller particles could contact the screen deck.

The particle velocity during the screening process has a huge impact on screen efficiency. The more particles that remain on the screen surface, the greater is the probability that these particles will pass through the screen deck. The average particle velocity for both screen decks is shown in Figure 11. The variation in the velocity for different material densities was not significant, and the higher-density particles had a slightly higher velocity. As Figure 11 shows, increasing the feed rate decreased the particle velocity along the screen deck as the number of collisions between the particles increased due to more material being on the screen deck.

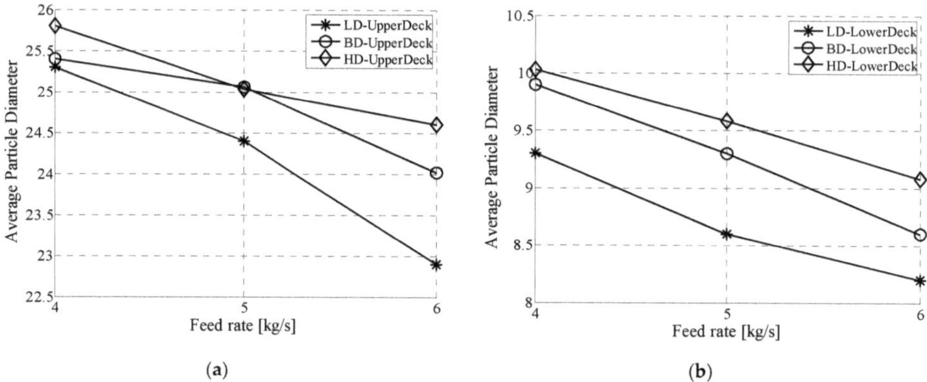

Figure 10. Average particle diameter in overflow material. (**a**) Upper deck. (**b**) Lower deck. (LD: Low-density material, BD: Between density material, and HD: High-density material)

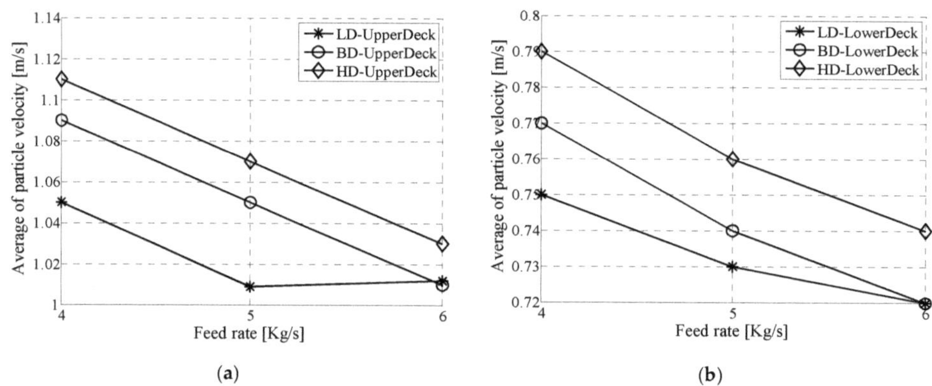

Figure 11. Average particle velocity for different material densities, (**a**) upper deck, (**b**) lower deck. (LD: Low-density material, BD: Between density material, and HD: High-density material)

Stratification can be affected by the difference in material densities; thus, a simulation with two different material densities was performed to determine the size of the effect of the material density on the screening performance. To generate the same volume of material for different densities, numbers of particles were used in the simulation instead of generating the materials by mass. As a result, the particle factory generated different numbers of particles to achieve a specified mass.

Figure 12 shows the total number of particles that passed through the screen deck in different sections. More high-density particles passed through the screen deck than for the low-density material. As Figure 12 shows, at the discharge point of the upper deck, there was almost the same number of particles for both the high- and low-density materials. However, the total number of low-density particles was greater at the discharge point in the lower deck.

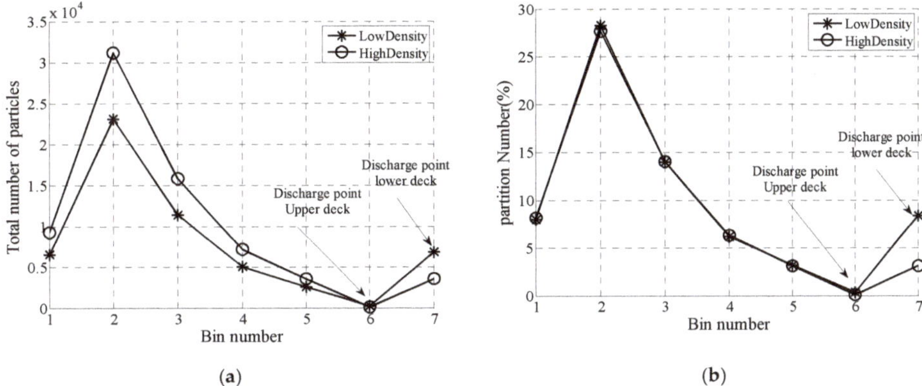

Figure 12. Total number of particles passed through screen deck in different sections with the feed for mixed particle densities. (**a**) Total number of particles. (**b**) Partition number.

Model Comparison

The mechanistic screen model presented by Soldinger [20] was used to study the theoretical passage rate along the screen deck. The factor β was calibrated to the value of 8.1, compared to 6.7 in Soldinger's original work [17]. A comparison was made by changing the material density of the incoming feed and estimating the passage rate at different positions in the same manner as used for the DEM model. The feed rate was fixed at 5 kg/s, and the size distribution was the same as the DEM simulation, to make the simulations comparable.

In Figure 13a, the result from model simulation and data from DEM are shown; Product 1 is the course product, Product 2 is the middle product, and Product 3 is the fine products at different intervals. As Figure 13a shows, the particle size distribution for the middle and fine products was around the same range. However, in the DEM simulation results, there was a significant portion of the middle size fraction, which remained in the coarse product fraction. This could be due to a few reasons; first, the shape of the particles defined in both simulations was different, since the particles in the DEM simulation had a more spherical shape, and the second, it could be due to the effect of friction. Friction between particles and between particles and the screen deck is different based on different rock materials used and screen deck material. In the case of the DEM model, this was covered by the Hertz–Mindlin contact model; in the mechanistic model, it was not explicitly formulated as an input variable.

The effect of different densities in the passage rate was studied, which can be seen in Figure 13b. The mechanistic model and the DEM simulation showed similar trends. The simulation result showed that the material with higher density had more of a chance to passage during the screening process into the first two bins.

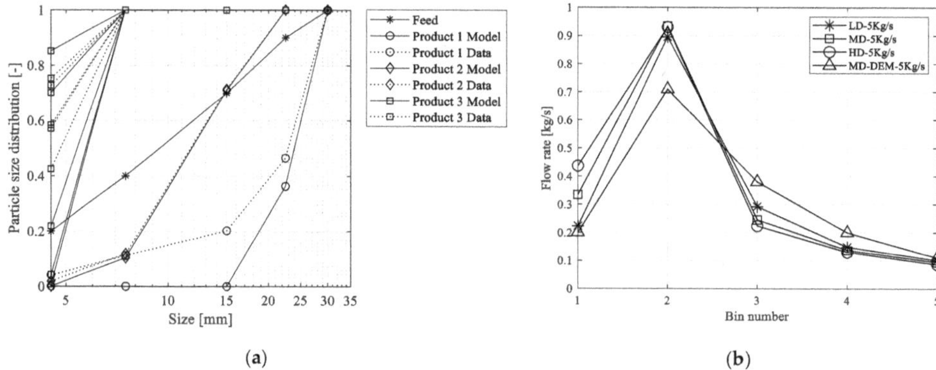

Figure 13. (**a**) Passage result from model simulation (product model), and DEM simulation (product data). (**b**) Passage rate in different screen sections by using screen model simulation.

4. Conclusions

Several main conclusions can be drawn from this study.

1. DEM is a powerful tool for calculating the overall efficiency and the product size distribution, while also enabling the analysis of important parameters, such as the size fraction (by sampling from any part of the screen); particle tracking; and the observation of bed material, which helps in stratification analysis. Therefore, DEM can provide a better understanding of different process parameters, such as how various particle densities have an effect on the stratification process and screen efficiency.
2. The increase in the passing percentage of small undersized particles mainly occurred at the upper deck, and as a result, the particle bed was thicker at the lower deck, which means that the stratification process mainly affects the screen efficiency in the lower deck.
3. In the simulations, the passage rate for the low-density material was lower than for the high-density material, since the low-density material had a lower stratification rate compared to the high-density material.
4. In the stratification of materials with various densities, it is easier for the high-density material to move vertically through the particle bed. Thus, the high-density material has a higher probability of passage.

5. Future work

In future work, experimental tests will be performed using a laboratory-scale vibrating screen that was used in all of the simulations in this paper.

Several parameters can be studied by using DEM, and experimental tests can be performed, such as determining the effect of the particle size distribution on the screen efficiency. The particle shape is an interesting factor that should be investigated to determine the significance of its effect on the stratification process.

Author Contributions: Conceptualization, A.D. and G.A.; methodology, A.D. and G.A.; software, A.D.; validation, A.D.; formal analysis, A.D.; investigation, A.D.; resources, A.D. and G.A; data curation A.D.; writing—original draft preparation, A.D.; writing—review and editing, G.A., E.H. and M.E.; visualization, A.D., G.A., E.H. and M.E.; supervision, G.A., E.H. and M.E.

Funding: The research presented in this paper was funded by Chalmers Area of Advance Production.

Acknowledgments: Thanks to all my colleagues at Chalmers Rock Processing research group for your support.

Conflicts of Interest: The authors declare no conflict of interest.

References

1. Wills, B.A. *Wills' Mineral Processing Technology: An Introduction to the Practical Aspects of Ore Treatment and Mineral Recovery*; Butterworth-Heinemann: Oxford, UK, 2011.
2. Asbjörnsson, G.; Bengtsson, M.; Hulthén, E.; Evertsson, M. Model of banana screen for robust performance. *Miner. Eng.* **2016**, *91*, 66–73. [CrossRef]
3. Soldinger, M. Influence of particle size and bed thickness on the screening process. *Miner. Eng.* **2000**, *13*, 297–312. [CrossRef]
4. Soldinger, M. Interrelation of stratification and passage in the screening process. *Miner. Eng.* **1999**, *12*, 497–516. [CrossRef]
5. Drahun, J.A.; Bridgwater, J. The mechanisms of free surface segregation. *Powder Technol.* **1983**, *36*, 39–53. [CrossRef]
6. Khakhar, D.V.; McCarthy, J.J.; Ottino, J.M. Radial segregation of granular mixtures in rotating cylinders. *Phys. Fluids* **1997**, *9*, 3600–3614. [CrossRef]
7. Bridgwater, J.; Foo, W.S.; Stephens, D.J. Particle mixing and segregation in failure zones—theory and experiment. *Powder Technol.* **1985**, *41*, 147–158. [CrossRef]
8. Wang, L.; Ding, Z.; Meng, S.; Zhao, H.; Song, H. Kinematics and dynamics of a particle on a non-simple harmonic vibrating screen. *Particuology* **2017**, *32*, 167–177. [CrossRef]
9. Clément, E.; Rajchenbach, J.; Duran, J. Mixing of a Granular Material in a Bidimensional Rotating Drum. *Europhys. Lett.* **1995**, *30*, 7–12. [CrossRef]
10. Xiao, H.; Umbanhowar, P.B.; Ottino, J.M.; Lueptow, R.M. Modelling density segregation in flowing bidisperse granular materials. *Proc. R. Soc. A Math. Phys. Eng. Sci.* **2016**, *472*, 20150856. [CrossRef]
11. Carvalho, R.; Tavares, L. A mechanistic model of SAG mills. In Proceedings of the IMPC 2014—27th International Mineral Processing Congress, Santiago, Chile, 20–24 October 2014.
12. Cleary, P.W. Ball motion, axial segregation and power consumption in a full scale two chamber cement mill. *Miner. Eng.* **2009**, *22*, 809–820. [CrossRef]
13. Chen, Y.-h.; Tong, X. Application of the DEM to screening process: A 3D simulation. *Min. Sci. Technol. (China)* **2009**, *19*, 493–497. [CrossRef]
14. Jahani, M.; Farzanegan, A.; Noaparast, M. Investigation of screening performance of banana screens using LIGGGHTS DEM solver. *Powder Technol.* **2015**, *283*, 32–47. [CrossRef]
15. Davoodi, A.; Bengtsson, M.; Hulthén, E.; Evertsson, C.M. Effects of screen decks' aperture shapes and materials on screening efficiency. *Miner. Eng.* **2019**, *139*, 105699. [CrossRef]
16. Cleary, P.W.; Sinnott, M.D.; Morrison, R.D. Separation performance of double deck banana screens—Part 1: Flow and separation for different accelerations. *Miner. Eng.* **2009**, *22*, 1218–1229. [CrossRef]
17. Teufelsbauer, H.; Wang, Y.; Chiou, M.; Wu, W. Flow–obstacle interaction in rapid granular avalanches: DEM simulation and comparison with experiment. *Granul. Matter* **2009**, *11*, 209–220. [CrossRef]
18. Yan, B.; Regueiro, R.; Sture, S. Three-dimensional ellipsoidal discrete element modeling of granular materials and its coupling with finite element facets. *Eng. Comput.* **2009**, *27*. [CrossRef]
19. Evertsson, J.Q.M. Framework for DEM Model Calibration and Validation. In Proceedings of the 14th European Symposium On Comminution And Classification, Gothenburg, Sweden, 7–11 September 2015.
20. Soldinger, M. Transport velocity of a crushed rock material bed on a screen. *Miner. Eng.* **2002**, *15*, 7–17. [CrossRef]

Article

The Effect of the Characteristics of the Partition Plate Unit on the Separating Process of −6 mm Fine Coal in the Compound Dry Separator

Changshuai Chen [1,2], Longlong Wang [1,2], Zhenfu Luo [1,2,*], Yuemin Zhao [1,2], Bo Lv [1,2], Yanhong Fu [1,2] and Xuan Xu [1,2]

[1] Key Laboratory of Coal Processing and Efficient Utilization of Ministry of Education, China University of Mining & Technology, Xuzhou 221116, China; cschen0608@163.com (C.C.); TS18040064A31@cumt.edu.cn (L.W.); ymzhao_paper@126.com (Y.Z.); lvbo2158@126.com (B.L.); tb18040021b4@cumt.edu.cn (Y.F.); xuxuansz@126.com (X.X.)

[2] School of Chemical Engineering and Technology, China University of Mining and Technology, Xuzhou 221116, China

[*] Correspondence: zfluo@cumt.edu.cn; Tel.: +86-13952197805

Received: 15 January 2019; Accepted: 30 March 2019; Published: 4 April 2019

Abstract: Although compound dry separation technology has been applied to industrial applications for +6 mm size fraction of coal separation, the technology has not been widely applied in the separation of fine coal (−6 mm). In this study, the effect of the partition plate unit characteristics on both the average density of particles in the bed uniformly and the final separation results in a fine coal separation process were studied. According to the results, the standard deviations of the corresponding density distribution were 0.08, 0.14, and 0.07 when the height of the partition plate was 2.5 cm, the partition plate angle was 35°, and the distance from the apex of partition plate on the backplane, was 12 cm, respectively, which were the lowest values at the same level. These results showed that the average density of particles in the bed was uniformly, and its corresponding density distribution contour map was more regular. When the amplitude was 2.8 mm, the frequency was 29 Hz, the height of the partition plate was 2.5 cm, the partition plate angle was 35°, and the distance from the apex of the partition plate to the backplane was 12 cm; as a result, the E value was 0.115 g/cm^3, the yield of the concentrate was 69.24%, the ash content was 12.52%, and the separation effect was better. The characteristics of the partition plate unit have an important effect on the separating process of −6 mm fine coal in the compound dry separator.

Keywords: compound dry separation; partition plate unit; bed density; fine coal

---

1. Introduction

Since coal is becoming the main source of energy in China, many people in China have focused on optimizing the use of the coal resource. It is reported that coal consumption accounts for approximately 60% of domestic energy consumption, and is expected to remain the main energy source in the future [1]. According to predictions [2], coal will remain the primary energy source until 2040 in China. However, effective utilization of coal is poor. The problems of environmental pollution, such as the generation of sulfur dioxide, acid rain, and soot, caused by the burning of coal that has not been beneficiated, are becoming increasingly serious. Therefore, developing more efficient processing technology for cleaning coal is very urgent [3]. It is also important to develop more efficient processing technology for cleaning coal to realize high efficiency, cleanliness, and comprehensive utilization of coal resources. With global climate change, the problems of global water shortages are becoming increasingly serious. According to statistics, the per capita share of water resources in China is only

1/17 of the world per capita share of water resources, with the per capita share of water resources in the northwest region being 1/2 of the national average. However, the distribution of coal resources is not matched with the distribution of water resources in China, with the coal reserves in the northwest of China accounting for 70% of the national coal deposits. At present, the main beneficiation methods for coal in China are wet and dry separation. Wet separation requires consumption of a large quantity of water resources, which is problematic given drought and water shortages in northwest China. Thus, it is very important to give careful consideration to the advantages of dry separation. Moreover, dry coal separation has more advantages compared to wet coal separation [4–11], for example, it does not consume water resources, does not increase the moisture content of separated products, and does not pollute groundwater and ecological environment. As a result, it can reduce damage to the environment. Moreover, dry separation technology has attracted the attention of the world because of its particular advantages, especially in arid and water-deficient regions.

Scholars and experts from all over the world have successively developed dry separation technologies (such as the wind jigging washing technology, the wind shaking bed technology, and the air heavy medium fluidized bed technology), some of which have been applied semi-industrially [12–21]. The compound dry separation technology is a successful separation technology currently applied in industry. Compared with the traditional wind separation process, the compound dry separation technology has the advantages of high separation efficiency, wide size fraction, large handling capacity, strong adaptability to external moisture of raw coal, and so on. The compound dry separation technology is mainly used for removing gangue from steaming coal, but its separation efficiency is not good for fine coal −6 mm. With the extensive application of modern machinery for coal mining, the proportion of −6 mm fine raw coal increases, even reaching as high as 70%. Thus, it is very important to study the compound dry separation technology. Various researchers [22–30] have studied the effects of parameters such as vibration frequency and amplitude, air velocity, and back angle on the separation performance of compound dry separators. These researchers have studied the influence of the structure of the compound dry separator on the separation effect, but not the characteristics of the partition plate.

This paper studied the effects of the characteristics of the partition plate unit on the separating process of −6 mm fine coal in the compound dry separator, and the separation efficiency of fine coal of −6 mm was improved. As a result, the compound dry separation could adapt to the high efficiency separation of raw coal with full particle size, and increase the productivity of −6 mm materials in the process of compound dry separation. The low ash content and the increase of −6 mm raw coal separation rate were of great significance to both the efficient separation of coal in arid and water-scarce areas of western of China as well as the effective separation of fine raw coal, which is also a supplement to the treatment of −6 mm fine raw coal.

2. Materials and Methods

2.1. Experimental Equipment

As shown in Figure 1, the experimental system of compound dry separation of fine coal is mainly composed of the following parts: a raw coal preparation part, an air supply dust removal part, and a separation part. The raw coal feeding device is composed of a buffer bin and a vibrating feeder. The raw coal feeding device can distribute the raw materials into the separation bed evenly. A vibration motor, hanging device, and trapezoidal separation bed constitute the main separating parts. A blower, dust collector, and draught fan constitute the air dust collection system. The blower provides the air power for separation that is fed into the separation bed evenly through the bed surface air distribution hole.

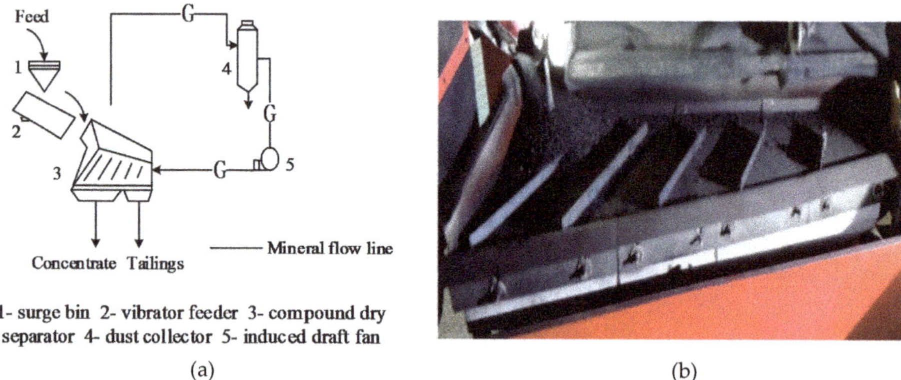

(a)   (b)

Figure 1. Flow chart of the model system of compound dry coal separation. (**a**) The schematic diagram; (**b**) the equipment diagram.

The materials are fed into the separating bed with a certain longitudinal and transverse inclinations by a vibrating feeder, and vibration from the vibration source acts on the bed surface. There are several air chambers under the bed which can control the air flow. The air is fed into the air chamber by a centrifugal ventilator. Through the air holes on the bed surface, the air acts on the separating materials upward. Under the double actions of vibration force and air fluidization force, the materials become loose and stratified according to density. The light materials are on the top of the bed surface and the heavy materials are on the bottom of the surface. Due to the inclined angle in the horizontal direction of the bed, the low-density materials slide down from the surface of the bed, so that the materials in the top layer of the bed surface are continuously discharged through the side discharge baffle and enters the concentrate section. The high-density materials gather at the bottom of the bed, and moves towards the gangue section under the action of the partition plate on the bed surface, and finally enters the tailings chute [28].

In this study, the different parameters of the height of the partition plate (H), the partition plate angle (θ) and the distance from the apex of the partition plate to the backplane (D) were tested. The best test parameters were found. Figure 2 is the three experimental parameters for the dry coal separation. The porosity of the bed surface is 65% and the bore diameter of bed surface is 3 mm.

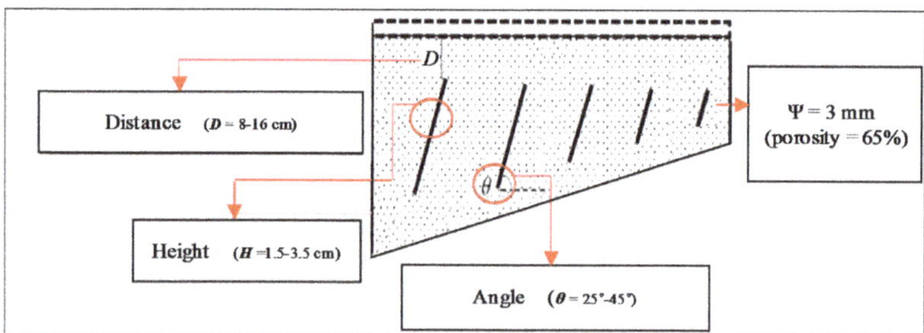

Figure 2. Three experiential parameters for the dry coal separation.

The separator parameters were listed in Table 1.

Table 1. Characteristics of testing system of fine compound dry separator.

Size (mm)	Operation Capacity (t/h)	Amplitude (mm)	Frequency (Hz)	Air Volume (m^3/h)
25–0	1–0.5	4–2	50–0	3517

2.2. Material Properties

The experimental sample was come from Inner Mongolia, which was produced from Bula mine in Zhongmeishan of Ordos. The raw coal used in this study was is brittle and muddy. The density distribution of the −6 + 1 mm size fraction was measured by the sink-float method, with results shown in Table 2.

Table 2. Results of sink-float test for −6 + 1 mm coal.

Density (g/cm^3)	Yield (%)	Ash Content (%)	Cumulative Float		Cumulative Sink	
			Yield (%)	Ash Content (%)	Yield (%)	Ash Content (%)
−1.40	17.92	5.51	17.92	5.51	100	25.24
1.4–1.5	31.01	7.42	48.93	6.72	82.08	29.55
1.5–1.6	18.63	14.11	67.56	8.76	51.07	42.99
1.6–1.7	7.23	21.15	74.79	9.95	32.44	59.56
1.7–1.8	4.32	34.81	79.11	11.31	25.21	70.58
1.8–2.0	3.35	54.58	82.46	13.07	20.89	77.97
2.00	17.54	82.44	100	25.24	17.54	82.44
Total	100	25.24				

Table 2 showed that the head ash of the coal was 25.24%. The yield of coal with density −1.6 g/cm^3 was nearly 68%, and the yield of density +1.8 g/cm^3 was greater than 20%.

2.3. Experimental Method

After the materials were fed into the bed surface from the feeder of the separator, the materials were forced to move due to the vibration force and air fluidization force of the bed surface. When the separator operated stably, samples were taken in the divided area of the bed surface to measure the ash content of the samples. When the separating process was over, the yield and ash content of the collecting concentrate and gangue were calculated.

During the all of the experiments, different parameters would be used when the characteristics of the partition plate unit were analysed, but the amplitude and frequency of the separator were a fixed value when the parameter of the partition plate was changed. The amplitude was 2.8 mm and the frequency was 29 Hz. When the height of the partition plate was changed to a range of 1.5 cm to 3.5 cm, the angle and distance were fixed. The angle was 35° and the distance from the apex of partition plate on the backplane was 12 cm. When the partition plate angle of the partition plate was changed to a range of 25° to 45°, the height was 2.5 cm, the distance from the apex of partition plate on the backplane was 12 cm. When the distance from the apex of the partition plate on the backplane was changed to a range of 8 cm to 16 cm, the height of partition plate was 2.5 cm, the partition plate angle was 35°.

2.4. Evaluation Index

In the actual separation process, there is not a suitable method to measure the bed surface density of the compound separating machine directly at the present stage, mainly the coal density calibration has more standards, such as true density and apparent density. No matter which kind of measuring method, when the amount of data is large, the measurement is more complicated. Therefore, a linear correlation curve of ash content and density is established in this experiment. The density of specific area of bed surface is calculated by sampling and measuring ash, and the density distribution of bed surface is obtained by interpolation function in the MATLAB software. The calculation formula of ash content-density was shown in Formula 1, and the correlation coefficient *R* was 0.9801, indicating that ash content was closely related to density in this experiment, and the ash-density correlation curve was shown in Figure 3.

$$y = 1.906 \times e^{[-((x-68.23)/109.9)^2]} \quad R = 0.9801 \tag{1}$$

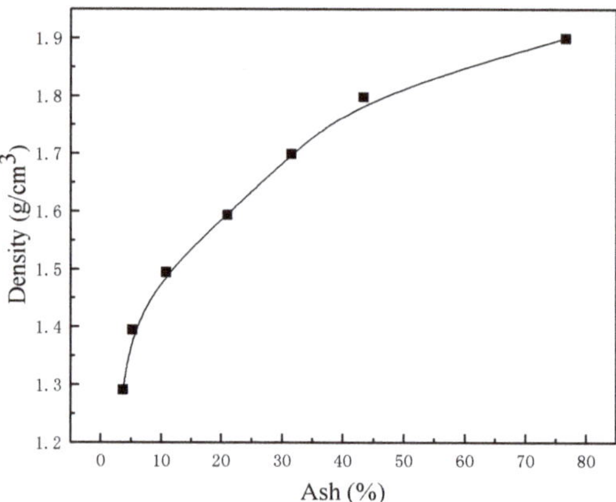

Figure 3. A schematic diagram of the linear correlation curve of ash density.

In order to understand the change of average density of particles in the bed during the process of compound dry separation of fine coal, the coordinate axis was established on the structure of trapezoidal bed surface, and the sampling was carried out in the separation process relying on the bed surface coordinates. The bed surface coordinates are shown in Figure 4. The *x*-axis is the direction from the feeding section to the gangue discharging section, and the *y*-axis is the direction from the feeding section to the discharging baffle. In the *x*-axis direction, the separation bed surface was divided into 10 sections, each segment spacing is 100 mm. In the *y*-axis direction, the separation bed surface was divided into 5 sections, each segment spacing is 100 mm. So, the bed surface was divided into many equal squares by *x*-axis and *y*-axis. In the direction of horizontal coordinate and longitudinal coordinate of bed surface, the areas which cross each other were sampling and measuring points. The sampling layout on the bed surface of the compound dry separator is shown in Table 3.

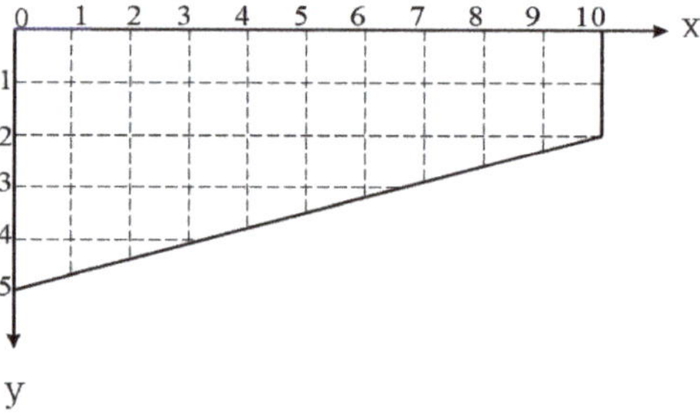

Figure 4. Schematic diagram of bed coordinates.

In each sampling process, after the equipment started to separate stably, the equipment was firstly turned off, and then the sampler was used to sample in the divided area on the bed surface, and all the samples in each area were taken away. Then the technology of division was acted on the sample, respectively. Finally, when the sample was suitable enough for ash content measurement, we tested the sample for ash content using a muffle furnace. we calculated the density through the ash content of the sample.

Table 3. Compound dry separation average density of particles in the bed test sampling layout table.

Base Point		Transverse Region									
		1	2	3	4	5	6	7	8	9	10
	1	1	2	3	4	5	6	7	8	9	10
Longitudinal Region	2	11	12	13	14	15	16	17	18	19	20
	3	21	22	23	24	25	26	27	28	29	30
	4	31	32	33	34	35	36	37			
	5	38	39	40							

(1) Standard deviation of the average density of particles in the bed distribution (S_ρ)

To characterize the uniformity of the average density of particles in the bed in a fixed area in the process of compound dry separation of fine coal, the density standard deviation (S_ρ) is used to reflect the degree of deviation from the average density of the measured area. The smaller is conducive to the separation of materials by density. The expression describing the data is as follows:

$$S_\rho = \sqrt{\frac{1}{n-1}\sum_{i=1}^{n}(\rho_i - \overline{\rho}_a)^2} \tag{2}$$

$$\overline{\rho}_a = \frac{1}{n}\sum_{i=1}^{n}\rho_i \tag{3}$$

In the formula, S_ρ is the standard deviation of the density distribution of the bed in units of g/cm^3. ρ_i is the value of density measurement at test point i in units of g/cm^3. $\overline{\rho}_a$ is the average value of density measurement in units of g/cm^3. n is the total number of measuring points.

(2) Comprehensive index K

The formula of the comprehensive evaluation index K is the yield of the concentrate divided by the ratio of ash reduction, which is the ration of concentrate ash to head ash. The expression of comprehensive index K is as follows:

$$K = \frac{\gamma_j}{A_j/A_y} \tag{4}$$

In the formula: γ_j is the yield of the concentrate in units of %. A_j is the ash content of concentrate in units of %. A_y is the ash content of raw coal in units of %.

Compared with other indices, the comprehensive index K not only covers the ash content and the yield of concentrate, but also takes the ash content of raw coal into account. The index can more comprehensively and accurately measure the effect of fine coal compound dry separation. A_j/A_y is the ratio of ash reduction, that is, the ratio of concentrate to raw coal ash content, which represents the effect of ash content reduction of the concentrate after separation. A higher K value denotes a better separation effect, that is, the lower ash content and the higher yield of concentrate.

3. Results and Discussion

3.1. Effect of the Height of the Partition Plate (H) on the Separation Process

Figure 5 showed the effect of the height of the partition plate (H) on the uniformity of the average density of particles in the bed and the separation efficiency. When the value of H was 2.5 cm, the density distribution of the bed materials was more uniform than that of the other two separation conditions. The isoline of density distribution was regular, and many low-density particles were accumulated in the concentrate section. Moreover, the high-density particles were distributed in the gangue section of the bed body. The value of S_ρ was 0.08, which was the lowest value under the same factors. This result indicated that the density stability of the bed surface for these parameters was uniform. When the value of H was 1.5 cm, the average density of particles in the bed distribution was less uniform, especially in the concentrate section and the middle coal section. One possible explanation for this result was that the value of H was too small, resulting in the bed materials accumulation thickness being too thin to form a more uniform density layer. Moreover, the thin bed made the materials between the bottom bed surface and the surface layer closing to the discharge section subject to the same force, which caused the surface low-density particles mixing with the bottom particles to move toward the gangue section, while the bottom materials moved toward the discharging section via the reverse thrust of the backplane. The value of S_ρ was 0.51, and the uniformity of average density of particles in the bed was poor. Figure 5 showed that, when the value of H was 3.5 cm, the density distribution of the bed was again less uniform, and some of the high-density particles were distributed in the middle coal section, the corresponding S_ρ was 0.35. The uniformity of the average density of particles in the bed was poor. Table 4 is S_ρ for different values of H.

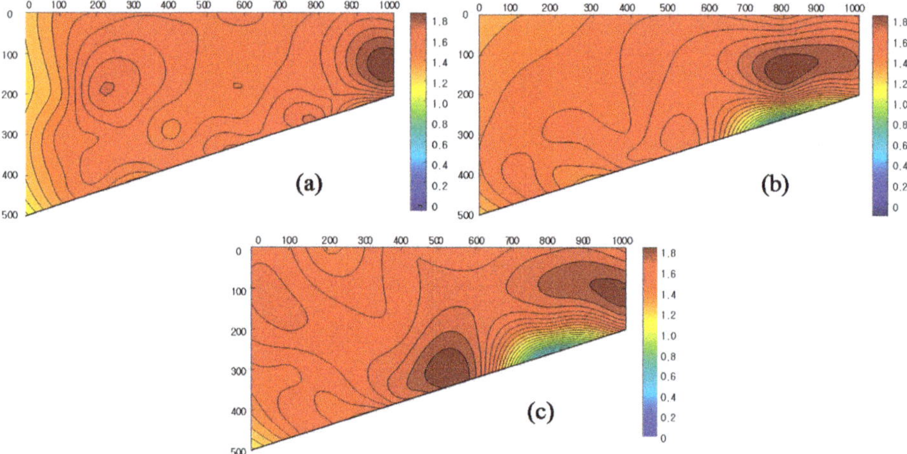

Figure 5. Effect of height of the partition plate on the average density of particles in the bed. (**a**) $H =$ 1.5 cm; (**b**) $H =$ 2.5 cm; (**c**) $H =$ 3.5 cm.

Table 4. S_ρ for different values of H.

H (cm)	S_ρ
1.5	0.51
2.5	0.08
3.5	0.35

Figure 6 showed that the effect of fine coal compound dry separation was directly related to H. When the value of H was 1.5 cm, the partition plate was so short that the materials in the bed were thin. Thus, it was difficult to form a stable density separating layer, causing the particles that have entered the bed to be discharged under the action of vibration force before stratification. As a result, the high-density materials were discharged with the concentrate section, and the yield of the concentrate was reduced. In the process of increasing H from 1.5 cm to 2.5 cm, the comprehensive index K increased gradually, and the separation effect became better, which showed the following: (1) the ash content of concentrate decreased sharply; and (2) the yield of the concentrate decreased little. When the value of H was 2.5 cm, although the yield of concentrate decreased slightly, the ash content and comprehensive index K of the concentrate reached the maximum value, and the separation effect was the best. When the H increased from 2.5 cm to 3.5 cm, the index K showed a downward tread. The yield of concentrate increased, and the ash content also increased linearly. The separation materials accumulating on the bed surface increased when the value of H was too high. The value of H not only affected the final separation effect but also affected the production efficiency of the separator to some extent.

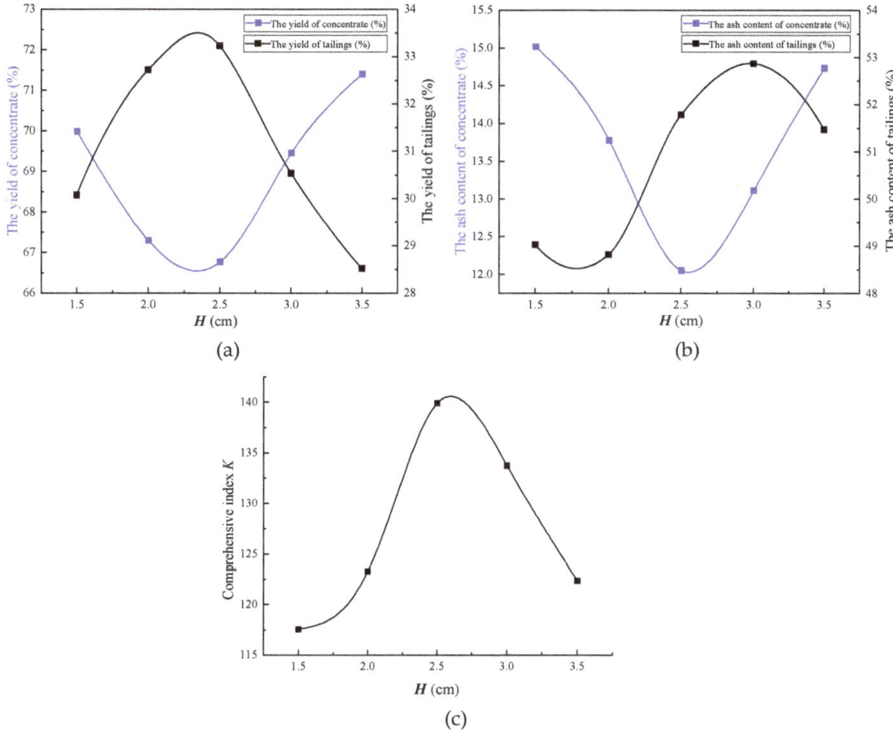

Figure 6. Effect of H on the separation effect. (**a**) Effect of the H on yield; (**b**) effect of the H on ash content; (**c**) effect of the H on K.

3.2. Effect of the Partition Plate Angle (θ) on the Separation Process

The effect of the partition plate on the uniformity of the average density of particles in the bed is shown in Figure 7. Figure 7a showed that, when the value of θ was 25°, the density distribution of bed surface was less uniform and had no obvious regularity, and the value of S_ρ was 0.37. The results showed that, in the separation process, the movement of particles was less uniform, and a uniform and stable density separation layer had not been formed in the bed. This phenomenon occurred because, when θ was small, the force acted on the particles, so that materials distributed along the backplane, thereby this fraction of the particles to be not sufficiently separated and discharged. Figure 7b showed that, when the value of θ was 35°, the isoline of the bed surface distribution presented a certain regularity, and the separation effect was better. The value of S_ρ reached 0.14, and the average density of particles in the bed value increased gradually in the x-axis direction. In the range of 0–500 mm of the x-axis, the density distribution of the bed surface was more uniform, which indicated that the average density distribution of bed surface in this area was uniform, and many low-density particles gathered in this area. In the range of 600–900 mm of the x-axis, the average density distribution of the bed increased obviously because many high-density materials moved to this side after separation. It showed that the particles in the bed surface could form a uniform average density layer when the value of θ was 35°, and the bed surface particles could be stratified according to the rule of density. When the value of θ was 45°, as shown in Figure 7c, there was no obvious regularity of the average density of particles in the bed distribution. In the range of 100–500 mm of the x-axis, the density distribution was less uniform than in the 35° case.

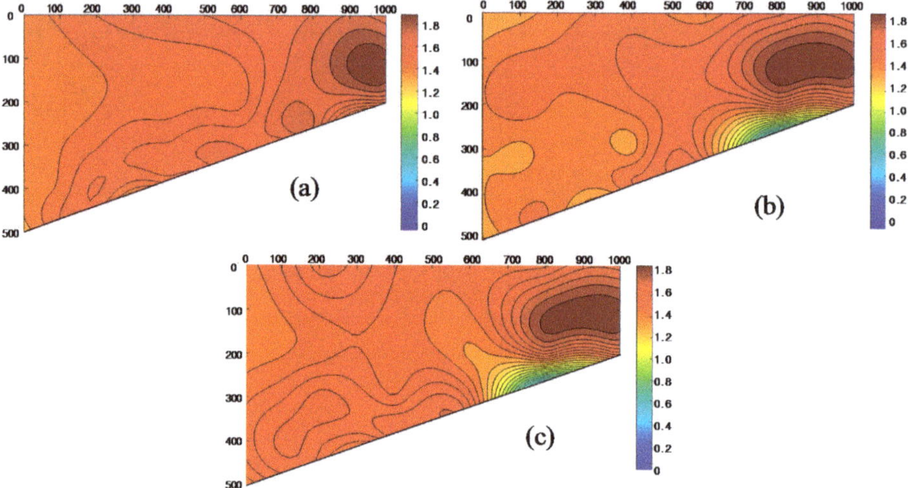

Figure 7. Effect of partition plate angle on the average density of particles in the bed. (**a**) $\theta = 25°$; (**b**) $\theta = 35°$; (**c**) $\theta = 45°$.

The effect of the partition plate angle on the separation is shown in Figure 8; when the value of θ increased from 25° to 35°, the yield of the concentrate increased quickly, and the ash content increased smoothly. The reason for these phenomena possible was that, when θ increased, the distance from the apex of partition plate to the backplane (D) reduced, resulting in the decrease of the tailings passage, thereby causing the concentrate carried by the tailings to be reduced. At the same time, the tailings passage became smaller, resulting in an increase in the amount of materials intercepted at the front of the separation bed surface. The materials intercepted on the bed surface increased, the density uniform separation layer could be formed under the actions of fluidization air, and the vibration. It made the high and low-density particles to separate quickly; thus, the separation efficiency was improved. When the value of θ was between 35° and 45°, the yield of the concentrate changed slowly, whereas the ash content increased sharply. The tailings passage on the bed surface became smaller, and the high-density particles could not be discharged through the tailings passage after separation, resulting in the deposition of high-density materials and the increase of the materials accumulation thickness in the upper part of the bed under the action of the reverse thrust of the backplane, followed by return of the materials to the separation bed. A large number of back-mixing particles after separation led to a sharp increase of the yield of the concentrate and an increase of the ash content of the concentrate. It could be also seen from the comprehensive index K that the range of K changed slowly in the process of increasing θ from 25° to 35°. When the value of θ further increased from 35°, the comprehensive index K decreased sharply; thus, the suitable value of θ was 35°. Table 5 is S_ρ for different values of θ.

Table 5. S_ρ for different values of θ.

θ (°)	S_ρ
25	0.37
35	0.14
45	0.61

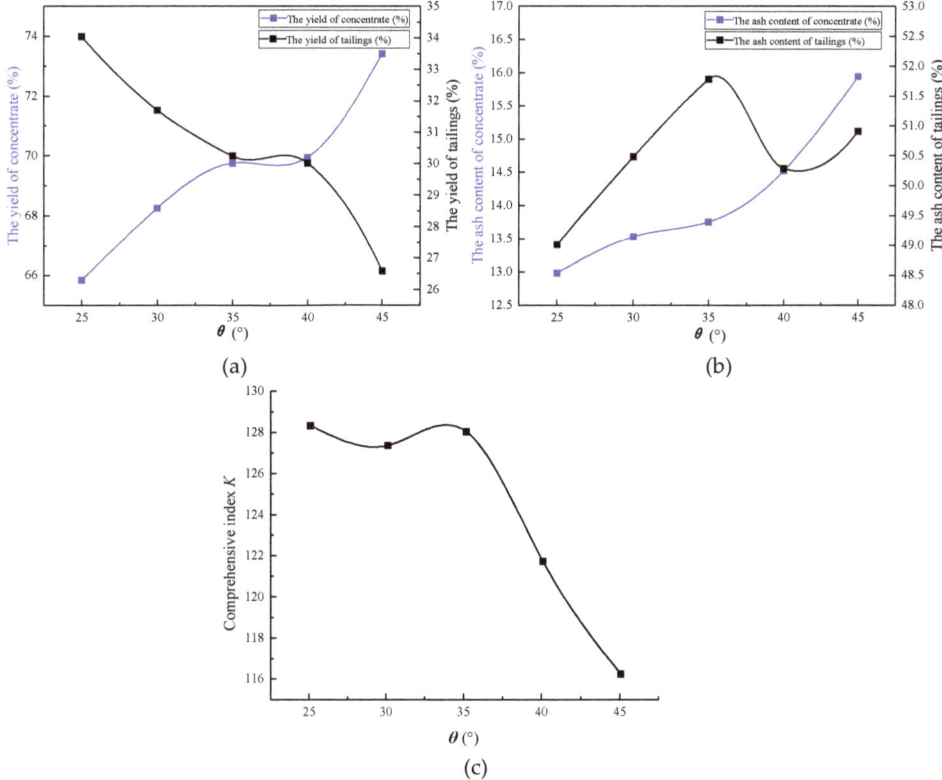

Figure 8. Effect of θ on the separation effect. (**a**) Effect of the θ on yield; (**b**) effect of the θ on the ash content; (**c**) effect of the θ on K.

3.3. Effect of the Distance from the Apex of the Partition Plate to the Backplane (D) on the Separation Process

The distance from the apex of the partition plate to the backplane (D) is the region between the apex of the partition plates and the D provides a channel to separate the coal on the bed surface. High-density tailings can easily pass when the D is large enough. Thus, D determines the moving speed of high-density materials which would be transported to the tailings section after separation. D also affects the handling capacity of the separator in a unit time. Figure 9 showed the influence of D on the stability of the average density of particles in the bed. When the value of D was 8 cm, as shown in Figure 9a, the distribution of average density of particles in the bed was poor, and the value of S_ρ was 0.45. A small number of high-density particles appeared in the 100–200 mm region of the x-axis because the value of D was small. It caused some high-density particles to spiral toward the backplane because the distance was so narrow that this fraction of particles could not be transported in time. As shown in Figure 9b, when the value of D was 12 cm, the average density of particles in the bed distribution showed a certain regularity, and the value of S_ρ reached the minimum of 0.07. In the range of 0–500 mm of the x-axis, the bed surface density distribution was more uniform, and the density was relatively small in this area. In this region, the materials could form layers, and the average density of particles in the bed was low. In the region of 500–700 mm of the x-axis, the density was relatively high because of the gradual movement of the high-density materials after separation from the concentrate section to this area. In the range of 700–900 mm of the x-axis, all the separated materials would be discharged from this section. Thus, in this area, the average density of particles in the bed was high. When the value of D was 16 cm, the density distribution of the bed surface was shown in Figure 9c. Although there were some regularities in the density distribution map, the average density of particles in the bed was low, the particles distribution on the bed surface was less uniform, and the corresponding value of S_ρ was 0.39. The uniformity of bed was poor. Table 6 is S_ρ for different values of D.

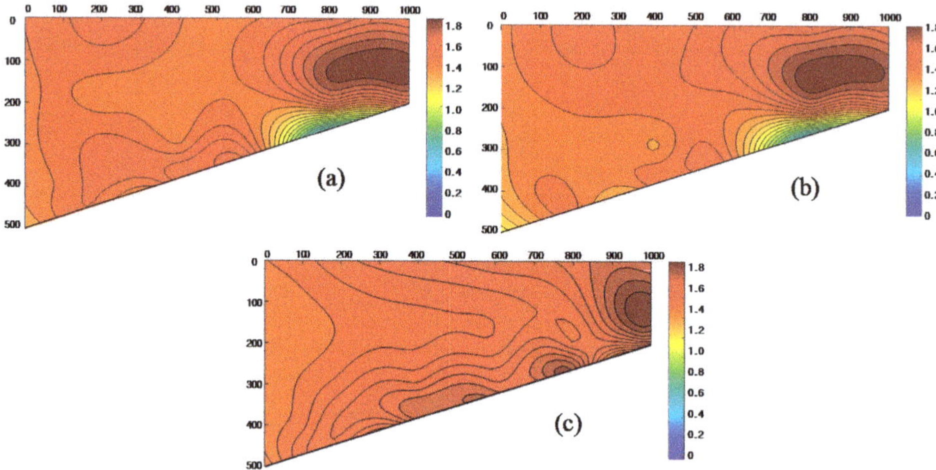

Figure 9. Effect of distance from the apex of partition plate to the backplane on the average density of particles in the bed. (**a**) $D = 8$ cm; (**b**) $D = 12$ cm; (**c**) $D = 16$ cm.

Table 6. S_ρ for different values of D.

D (cm)	S_ρ
8	0.45
12	0.07
16	0.39

The effect of the change in D on the separation of fine coal was shown in Figure 10. The diagram showed that the separation effect of fine coal was affected directly by the change in the value of D. Overall, the separation effect of compound dry separation improved when the distance increased gradually. However, when the value of D exceeded 12 cm, the separation effect deteriorated sharply. Figure 10a showed that, during the process of increasing D from 8 cm to 12 cm, the ash content of the concentrate decreased sharply and the yield of concentrate decreased slowly. Moreover, the comprehensive index K reached the maximum when the value of D was 12 cm. The results showed that the separation effect of fine coal was the best. One possible explanation for this result was that when the value of D was small, the high-density materials at the bottom of the bed after stratification moved near the backplane under the excitation force of the bed surface. Because the value of D was too small, this part of the stratified high-density particles could not be moved to the tailings section in time. In contrast, these particles returned to the concentrate section of the separation bed under the action of backplane reverse thrust. Furthermore, some high-density particles would be intercepted when the D was smaller. This fraction of back-mixing particles would aggravate the fluctuation of the average density of particles in the bed to a certain extent and worsen the overall separating effect. Under the combined action of the two factors, the yield and ash content of the concentrate were relatively high. When the value of the D was 12 cm, the yield of concentrate decreased sharply, the ash content increased linearly, and the separation index became the worst.

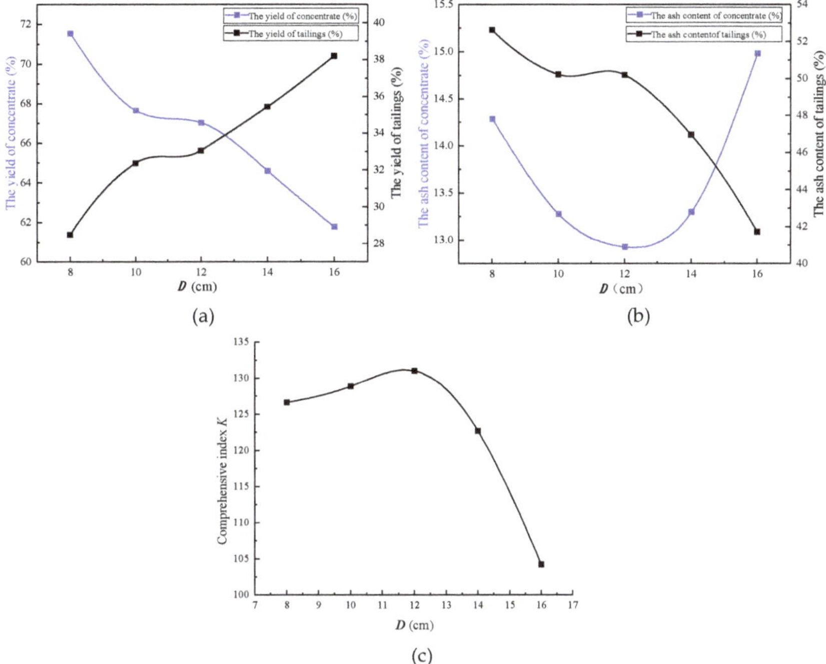

Figure 10. Effect of D on the separation effect. (**a**) Effect of the D on yield; (**b**) effect of the D on ash content; (**c**) effect of the D on K.

3.4. Determination of the Optimum Partition Parameter

In this paper, the Box-Behnken response surface method of Design-Expert software was used to analyze the influence of the operation factors on the separating effect. The operating factors were the following: amplitude, frequency, height of the partition plate, partition plate angle, and distance from the apex of the partition plate to the backplane. The basic parameters in the separation experiments were shown in Table 7, with each factor set at three levels. The selected amplitude was 2.8 mm, the frequency was 29 Hz, the H was 2.5 cm, the θ was 35°, and the D was 12 cm. The same numerical interval between the left and right sides were taken to carry out the multi-factor experimental design with these five data as the center. The reason why these ranges of variation were chosen was that we just chose the optimal parameters under the operability of the machine. The recommended model was the quadratic polynomial model (Prob > F, value of Prob was 0.0075, indicating that the model was significant. The Prob is the possibility). The analysis results are shown in Table 8.

It can be seen in Table 7 that the mathematical model is expressed as follows:

(1) Mathematical model in terms of factors of code:

$$K = 127.4944 + 2.48A + 0.25B + 2.02C - 0.33D - 2.03E - 0.60AB - 0.61AC - 1.12AD +$$
$$9.51AE - 0.17BC + 7.46BD + 0.70BE + 4.78CE + 6.82DE - 7.70A^2 - 6.41B^2 \tag{5}$$
$$- 9.52C^2 - 8.96D^2 - 8.92E^2$$

(2) Mathematical model in terms of actual operational factors:

$$K = -5358.55 + 946.63\alpha + 324.61\beta + 163.12\gamma - 23.35\delta - 60.02\varepsilon - 3.0\alpha\beta - 6.13\alpha\gamma$$
$$- 1.12\alpha\delta + 23.78\alpha\varepsilon - 0.35\beta\gamma + 1.49\beta\delta + 0.35\beta\varepsilon + 0.04\gamma\delta + 4.78\gamma\varepsilon + 0.68\delta\varepsilon - 192.48\alpha^2 \tag{6}$$
$$- 6.41\beta^2 - 32.08\gamma^2 - 0.36\delta^2 - 2.23\varepsilon^2$$

In this equation, α, β, γ, δ, ε represented amplitude, frequency, height, angle and distance, respectively.

Table 7. Basic operation parameters in the separation experiments.

Code	Factors	Unit	Minimum	Maximum	Minimum Code	Maximum Code
A	Amplitude	mm	2.6	3.0	−1	1
B	Frequency	Hz	28	30	−1	1
C	Height	cm	2.0	3.0	−1	1
D	Angle	°	30	40	−1	1
E	Distance	cm	10	14	−1	1

Table 8. Variance analysis of quadratic models.

Source	Sum of Squares	df	Mean Square	F value	Prob > F
Model	2618.31	20.00	130.92	3.47	0.00
A-Amplitude	98.70	1.00	98.70	2.62	0.12
B-Frequency	1.01	1.00	1.01	0.03	0.87
C-Height	65.55	1.00	65.55	1.74	0.20
D-Angle	1.78	1.00	1.78	0.05	0.83
E-Distance	65.77	1.00	65.77	1.74	0.20
AB	1.44	1.00	1.44	0.04	0.85
AC	1.50	1.00	1.50	0.04	0.84
AD	5.04	1.00	5.04	0.13	0.72
AE	361.76	1.00	361.76	9.60	0.00
BC	0.12	1.00	0.12	0.00	0.96
BD	222.61	1.00	222.61	5.90	0.02
BE	1.97	1.00	1.97	0.05	0.82
CD	0.03	1.00	0.03	0.00	0.98
CE	91.30	1.00	91.30	2.42	0.13
DE	185.78	1.00	185.78	4.93	0.04
Residual	942.53	25.00	37.70		
Lack of fit	942.53	20.00	47.13		
Pure error	0.00	5.00	0.00		

Figure 11 showed the comprehensive index K response surface and its contour diagram. It could be seen from Figure 11a that the variation trend of comprehensive index K along the amplitude direction was more obvious than that along the D direction. From the contour map mapped by the response surface, the sensitivity of comprehensive index K to amplitude change was found to be more obvious than the variation of D. This result indicated that amplitude change had a great influence on the final separating efficiency. Figure 11b showed a comprehensive index K response surface and its contour under the interaction of different frequencies and angles. The response surface curve revealed that the trend of comprehensive index K along the direction of partition plate angle was steeper than that along the direction of amplitude. The corresponding contour line also showed that the comprehensive index K was more sensitive to the change of θ. This result showed that, in the separation process, θ had a greater effect on the final separating efficiency than the amplitude. Figure 11c showed a comprehensive index K response surface and its contour line under the influences of different values of H and D. The response surface curve revealed that the variation trend of comprehensive index K along the H direction was steeper than that along the D direction. The plot showed that the change of D had less influence on the comprehensive index K than the change of H. The contour of the corresponding surface also showed that the comprehensive index K was more sensitive to the change of H. It could be seen from the Table 8 that the significant factors that influence the comprehensive index K were ranked in level of influence as follows: amplitude > H > D > θ > frequency. Table 9 is the separation results of fine coal in a compound dry separator, Table 10 presents the partition coefficient results of −6 + 1 mm fine coal for compound dry separation and Figure 12 is the partition coefficient of the compound dry separation −6 + 1 mm fine coal.

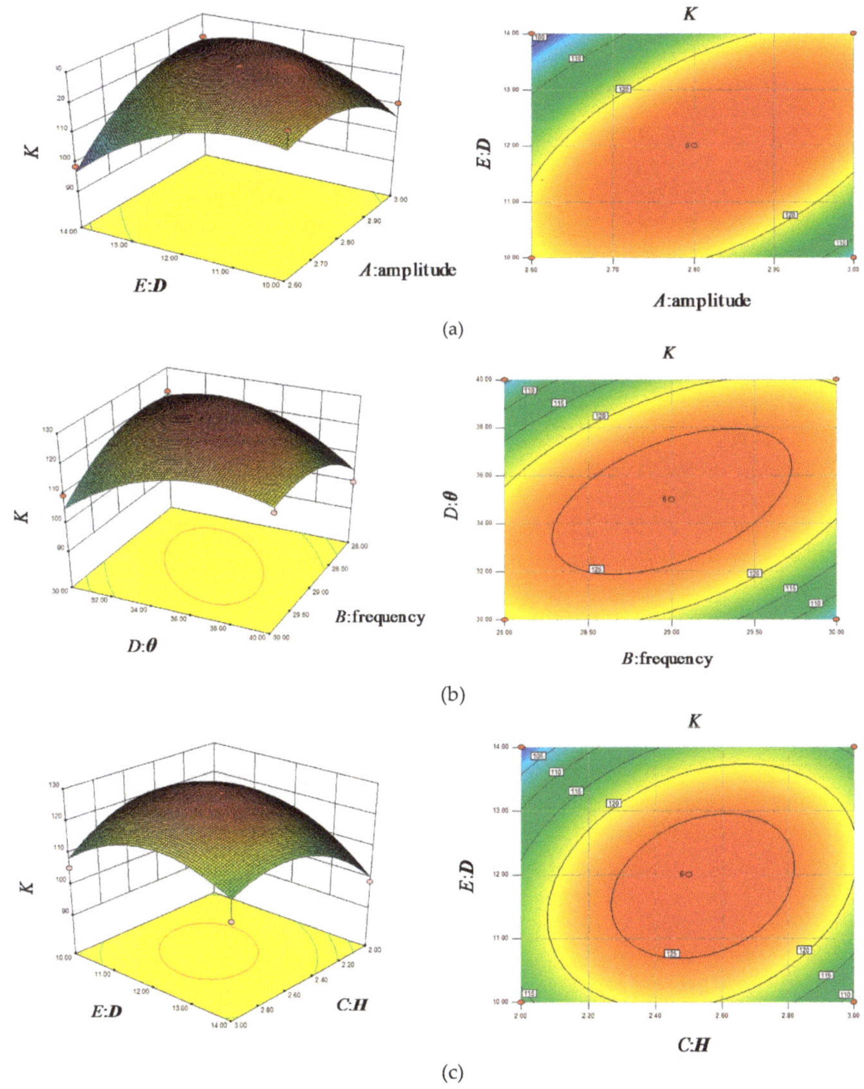

Figure 11. Response surface of *K* of −6 + 1 mm fine coal for various factors. (**a**) The response surface and its contour of *K* for different values of the amplitudes and *D*; (**b**) the response surface and its contour of *K* for different values of the frequencies and θ; (**c**) the response surface and its contour of *K* under different *H* and *D*.

Table 9. Separation results of fine coal in a compound dry separator.

Separation Text	Ash Content of Raw Coal (%)	Concentrate		Tailings	
		Yield (%)	Ash Content (%)	Yield (%)	Ash Content (%)
Results	25.24	69.24	12.52	30.76	53.87

Table 10. Partition coefficient results of −6 + 1 mm fine coal for compound dry separation.

Density (g/cm^3)	Average Density (g/cm^3)	Feed Density Distribution (%)	Tailings Sink-Float Results (%)		Concentrate Sink-Float Results (%)		Calculated Feedstock Sink-Float Results (%)	Partition Coefficient (%)
			Of Products (%)	Of Feedstock (%)	Of Products (%)	Of Feedstock (%)		
−1.4	1.30	17.92	0.76	0.23	25.63	17.75	17.98	1.29
1.4–1.5	1.45	31.01	7.62	2.34	40.50	28.04	30.39	7.71
1.5–1.6	1.55	18.63	14.35	4.41	21.73	15.05	19.46	22.68
1.6–1.7	1.65	7.23	7.11	2.19	7.33	5.08	7.27	30.11
1.7–1.8	1.75	4.32	9.93	3.05	1.83	1.27	4.32	70.70
1.8–2.0	1.90	3.35	8.48	2.61	0.63	0.43	3.04	85.72
+2.00	2.00	17.54	51.76	15.92	2.35	1.63	17.55	90.73
Total		100.00	100.00	30.76	100.00	69.24	100.00	

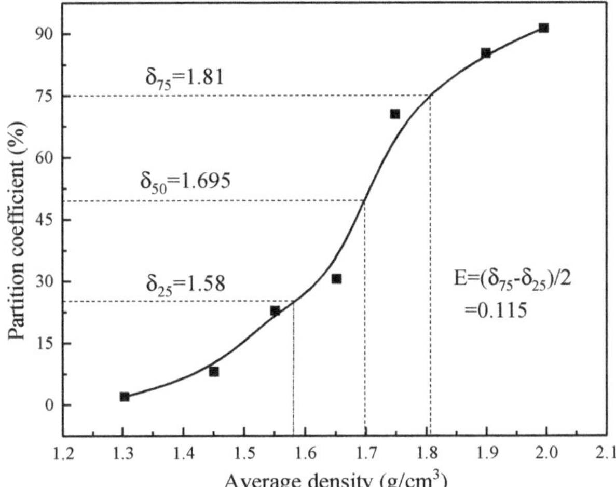

Figure 12. Partition coefficient of the compound dry separation −6 + 1 mm fine coal.

After the parameters were optimized by multi-factor test, the −6 + 1 mm fine coal compound dry separation test was conducted. The experimental results were as follows. By drawing the partition curve to calculate the E_p value, the value was 0.115, and the yield of concentrate was 69.24%. The fine coal ash content dropped to 12.52 %.

4. Conclusions

(1) The effects of the characteristics of the partition plate element on the separation of fine coal were studied. For the height of the partition plate (H) is 2.5 cm, the partition plate angle (θ) is 35°, and the distance from the apex of partition plate to the backplane (D) is 12 cm, the standard deviation (S_ρ) of the corresponding density were obtained as 0.08, 0.14, 0.07, respectively, reaching the lowest value under the same factors. This finding showed the following: (a) the density of the bed surface was more uniformity; and (b) the regularity of the isoline of the average density of particles in the bed distribution was obvious, that is, the density values of the concentrate section and the middle coal section to the tailings section increased gradually and the materials on the bed surface were separated according to the density distribution.

(2) Using Design-Expert software and univariate test factor, the test results showed that, under the conditions of the amplitude of 2.8 mm, the frequency of 29 Hz, the height of the partition plate (H) of 2.5 cm, the partition plate angle (θ) of 35°, and the distance from the apex of the partition plate to the backplane (D) of 12 cm, both the comprehensive index K and the separation efficiency of fine coal were the best. Through the numerical analysis of the response surface design experiment, it was concluded that the ranking of the effect of each factor on the comprehensive index K was: amplitude > $H > D > \theta >$ frequency.

(3) Using the optimum set of operating conditions, the separation test of −6+1 mm fine coal showed that the ash content of raw coal was 25.24%, the yield of concentrate was 69.24%, the separated coal ash was divided into 12.52%, the probable E_p was 0.115 g/cm^3.

Author Contributions: Data curation, C.C., L.W.; Methodology, Z.L., C.C.; Software, C.C., B.L., Y.F.; Writing-review and editing, C.C., X.X.; Validation, Z.L., Y.Z.; Writing—original draft, C.C., L.W.; Formal analysis, C.C., B.L., Y.F.; Supervision, Z.L., Y.Z.; Project administration, Z.L.

Funding: The authors acknowledge the financial support by Fundamental Research Funds for the Central Universities (2018XKQYMS05).

Acknowledgments: The author appreciates the teacher for his careful guidance and for the help of his brother and sister.

Conflicts of Interest: The authors declare no conflict of interest.

References

1. Circular of the National Energy: Administration on the Issuance of Guidance on Energy Work in 2018. National Energy Administration, 2017; pp. 2–26. Available online: http://zfxxgk.nea.gov.cn/auto82/201803/t20180307_3125.htm (accessed on 3 April 2019). (In Chinese)
2. Statistical Bulletin of National Economic and Social Development for 2018. National Bureau of Statistics of the people's Republic of China, 2018; pp. 2–28. Available online: http://www.stats.gov.cn/tjsj/zxfb/201902/t20190228_1651265.html (accessed on 3 April 2019). (In Chinese)
3. Lu, X.; Yu, Z.F.; Wu, L.X.; Yu, J.; Chen, G.F.; Fan, M.H. Policy study on development and utilization of clean coal technology in China. *Fuel Process. Technol.* **2008**, *89*, 475–484. [CrossRef]
4. Yang, X.L.; Zhao, Y.M.; Luo, Z.F. Dry cleaning of fine coal based on gas-solid two-phase flow-a review. *Chem. Eng. Technol.* **2017**, *40*, 439–449. [CrossRef]
5. Zhao, Y.M.; Tang, L.G.; Luo, Z.F.; Liang, C.C.; Xing, H.B.; Duan, C.L.; Song, S.L. Fluidization characteristics of a fine magnetite powder fluidized bed for density-based dry separation of coal. *Sep. Sci. Technol.* **2012**, *47*, 2256–2261.
6. Yang, X.L.; Zhao, Y.M.; Luo, Z.F.; Song, S.L.; Duan, C.L.; Dong, L. Fine coal dry cleaning using a vibrated gas-fluidized bed. *Fuel Process. Technol.* **2013**, *106*, 338–343. [CrossRef]
7. Yang, X.L.; Zhao, Y.M.; Luo, Z.F.; Song, S.L.; Chen, Z.Q. Fine coal dry beneficiation using autogenous medium in a vibrated fluidized bed. *Int. J. Miner. Process.* **2013**, *125*, 86–91. [CrossRef]
8. Luo, Z.F.; Tang, L.G.; Dai, N.N.; Zhao, Y.M. The effect of a secondary gas-distribution layer on the fluidization characteristics of a fluidized bed used for dry coal beneficiation. *Int. J. Miner. Process.* **2013**, *118*, 28–33. [CrossRef]
9. Dong, L.; Zhao, Y.M.; Xie, W.N.; Duan, C.L.; Li, H.; Hua, C.P. Insights in active pulsing air separation technology for coarse coal slime by DEM-CFD approach. *J. Cent. South Univ.* **2013**, *20*, 3660–3666. [CrossRef]
10. Zhao, Y.M.; Li, G.M.; Luo, Z.F.; Zhang, B.; Dong, L. Industrial application of a modularized dry-coal-beneficiation technique based on a novel air dense medium fluidized bed. *Int. J. Coal Prep. Util.* **2016**. [CrossRef]
11. Dong, L.; Zhao, Y.M.; Peng, L.P.; Zhao, J.; Luo, Z.F.; Liu, Q.X.; Duan, C.L. Characteristics of pressure fluctuations and fine coal preparation in gas-vibro fluidized bed. *Particuology* **2015**, *21*, 146–153. [CrossRef]
12. Zhang, B.; Zhu, G.Q.; Lv, B.; Yan, G.H. A novel and effective method for coal slime reduction of thermal coal processing. *J. Clean. Prod.* **2018**, *198*, 19–23. [CrossRef]
13. Chao, N.I.; Yuan, X.G.; Gui, J.Z.; Bo, L.; Li, P.Y. Problem analysis and optimization test for "2+2" coal slime separation process. *J. China Coal Soc.* **2013**, *38*, 2035–2041.
14. Luo, Z.F.; Zhao, Y.M.; Chen, Q.R.; Tao, X.X.; Fan, M.M. Research on gas distribution of dense phase high density fluidized bed for separation. *J. China Univ. Min. Technol.* **2004**, *33*, 237–240.
15. Luo, Z.F.; Zhao, Y.M.; Fan, M.M.; Tao, X.X.; Chen, Q.R. Density calculation of a compound medium solids fluidized bed for coal separation. *J. South. Afr. Inst. Min. Metall.* **2006**, *106*, 749–752.
16. Dwari, R.K.; Rao, K.H. Dry beneficiation of coal-A review. *Miner. Process. Extr. Metall. Rev.* **2007**, *28*, 177–234. [CrossRef]
17. Tang, L.G.; Zhao, Y.M.; Luo, Z.F.; Liang, C.C.; Chen, Z.Q.; Xing, H.B. The Effect of fine coal particles on the performance of gas-solid fluidized beds. *Coal Prep.* **2009**, *29*, 265–278. [CrossRef]
18. Dong, H.L.; Liu, C.S.; Zhao, Y.M.; Zhao, L.L. Review of the development of dry coal preparation theory and equipment of the development of dry coal preparation theory and equipment. *Adv. Mater. Res.* **2013**, *619*, 239–243. [CrossRef]
19. Zhao, Y.M.; Luo, Z.F.; Chen, Z.Q.; Tang, L.G.; Wang, H.F.; Xing, H.B. The effect of feed-coal particle size on the separating characteristics of a gas-solid fluidized bed. *J. South Afr. Inst. Min. Metall.* **2010**, *110*, 219–224.
20. Luo, Z.F.; Zuo, W.; Tang, L.G.; Zhao, Y.M.; Fan, M.M. Preparation of solid medium for use in separation with gas-solid fluidized beds. *Min. Sci. Technol.* **2010**, *20*, 743–746. [CrossRef]
21. Zhang, B.; Luo, Z.F.; Zhao, Y.M.; Lv, B.; Song, S.L.; Duan, C.L.; Chen, Z.Q. Effect of a high-density coarse-particle layer on the stability of a gas-solid fluidized bed for dry coal beneficiation. *Int. J. Miner. Process.* **2014**, *132*, 8–16. [CrossRef]

22. Yu, X.D.; Luo, Z.F.; Li, H.; Yang, X.; Cai, L. Effect of vibration on the separation efficiency of high-sulfur coal in a compound dry separator. *Int. J. Miner. Process.* **2013**, *157*, 195–204. [CrossRef]
23. Yu, X.D.; Luo, Z.F.; Li, H.B.; Yang, X.L.; Zhou, E.H.; Jiang, H.S.; Wu, J.D.; Song, S.L.; Cai, L.H. Effect of vibration on the separation efficiency of high-sulfur coal in a compound dry separator. *Int. J. Miner. Process.* **2016**, *157*, 195–204. [CrossRef]
24. Yu, X.D.; Luo, Z.F.; Li, H.B.; Yang, X.L. Separation of <6 mm oil shale using a compound dry separator. *Sep. Sci. Technol.* **2017**, *52*, 1615–1623.
25. Yu, X.D.; Luo, Z.F.; Li, H.B.; Gan, D.Q. Effect of vibration on the separation efficiency of oil shale in a compound dry separator. *Fuel* **2018**, *214*, 242–253. [CrossRef]
26. Ling, X.Y.; He, Y.Q.; Zhao, Y.M.; Li, G.M.; Wang, J.; Wen, B.F. Distributions of size, ash, and density of coal particles along the front discharge section of a compound dry separator. *Int. J. Coal Prep. Util.* **2017**, *38*, 422–432. [CrossRef]
27. Zhang, B.; Akbari, H.; Yang, F.; Mohanty, M.K.; Hirschi, J. Performance optimization of the FGX dry separator for cleaning high-sulfur coal. *Int. J. Coal Prep. Util.* **2001**, *31*, 161–186. [CrossRef]
28. Ling, X.Y.; He, Y.Q.; Li, G.M.; Tang, X.L.; Xie, W.N. Separation performances of different particle sizes using an industrial FGX dry separator. *Int. J. Coal Prep. Util.* **2016**. [CrossRef]
29. Li, H.B.; Luo, Z.F.; Zhao, Y.M.; Wu, W.C.; Zhang, C.Y.; Dai, N.N. Cleaning of South African coal using a compound dry cleaning apparatus. *Min. Sci. Technol. (China)* **2011**, *21*, 117–121.
30. Yang, X.L.; Zhao, Y.M.; Li, G.M.; Luo, Z.F.; Chen, Z.Q.; Liang, C.C. Establishment and evaluation of a united dry coal beneficiation system. *Int. J. Coal Prep. Util.* **2012**, *32*, 95–102.

Article

Scrubbing and Inhibiting Coagulation Effect on the Purification of Natural Powder Quartz

Xin Du, Chao Liang, Donglai Hou, Zhiming Sun * and Shuilin Zheng *

School of Chemical and Environmental Engineering, China University of Mining and Technology (Beijing), Beijing 100083, China; dxcumtb@163.com (X.D.); lccumtb@163.com (C.L.); hdlcumtb@163.com (D.H.)
* Correspondence: zhimingsun@cumtb.edu.cn (Z.S.); zhengsl@cumtb.edu.cn (S.Z.); Tel.: +86-10-62339920 (Z.S.)

Received: 27 January 2019; Accepted: 19 February 2019; Published: 26 February 2019

Abstract: The low removal efficiency of fine clay impurities in natural powder quartz (NPQ) is the main problem that affects the practical utilization of this natural resource. In this work, detailed characterizations of NPQ and clay impurities in NPQ were analyzed by SEM-EDS, mineral liberation analysis (MLA) and impurities distribution analysis. A combined physical purification process, including sieving, scrubbing and centrifugation, was applied to remove the clay impurities. It was observed that the fine clay impurities adhering on quartz surface were effectively liberated by scrubbing, and the content of Fe_2O_3 and Al_2O_3 in the concentrate decreased from 0.48% and 0.40% to less than 0.01% and 0.02% at pH 9.3 or when the dosage of sodium hexametaphosphate (SHMP) was 1×10^3 g/t. The coagulation interaction between quartz and impurities including hematite and orthoclase were analyzed based on the classical Deyaguin-Landau-Verwey-Overbeek (DLVO) theory. The results indicated that the main coagulation affecting the separation efficiency was the heterocoagulation between quartz and impurities and homocoagulation among hematite particles. Furthermore, adding regulators such as sodium hydroxide (NaOH) or SHMP could significantly decrease the zeta potential of minerals and thus increase the total interaction energy (V_T), which could effectively improve the dispersion of these fine impurity particles, and consequently improve the removal efficiency of impurities. Reverse increase of the zeta potential of minerals in strongly alkaline solutions or excessive SHMP were detected, which was likely the main factor limiting the further improvement of the purification efficiency.

Keywords: natural powder quartz; scrubbing; centrifugation; DLVO theory; coagulation

---

1. Introduction

Quartz is a highly critical and strategic material for advanced and emerging industrial technologies, which can be used extensively in advanced manufacturing of materials such as optical fibers, semi-conduct materials, highly pure glass in aerospace, etc. [1,2]. Considering the constant demand for better living standards of the increasing population, more and more quartz resources will be required in the future. Among different natural quartz resources, crystal is still the first choice for high purity quartz preparation. However, high purity crystal resources are rare in the world. Hence, researchers have tried to purify other natural quartz resources such as quartzite, quartz sand and white sand as substitutes [3,4]. In order to remove impurities, various processing methods were used. Physical methods such as magnetic separation, reverse flotation [5,6], and chemical treatment such as acidic leaching by hydrogen fluoride (HF) combined with other different fractions strong industrial acids such as HCl, HNO_3 or H_2SO_4 [2,7] were usually employed. In addition, some other new purification methods were also proposed, including bioleaching, supersonic-enhanced leaching and microwave explosion [8,9]. Because of the scarcity of high-quality resources and the high processing costs, developing feasible and economical processing technology is especially important.

NPQ is a natural quartzose material which is widely distributed in China; it is formed by the the weathering of quartzite [10]. The characteristics of this resource include about 97–99% SiO_2, integrity granular shape and extremely fine grain (D_{90} is approximately 67 µm and D_{50} is around 39 µm) compared with other quartz resource. However, this resource is contaminated by different contents of impurities such as Fe_2O_3 and Al_2O_3 coming from fine clay minerals, which might significantly limit its application. Additionally, these fine clay impurities generated in the weathering course or exploitation are difficult to remove.

The traditional purification methods for NPQ include elutriation, flushing by quantitative water and spontaneous precipitation in a settling basin [11]. However, these methods have low separating efficiencies and high operation costs in practical applications. With the improvement in separation technologies, some methods such as magnetic separation and reverse flotation were applied for the purification of NPQ [12,13]. However, all these reported methods were inefficient for the removal of fine clay minerals in NPQ. The main reason for the low removal efficiency is that the impurity contents were so scarce that they could not be definitively distinguished by traditional methods, which limited advances in purification methods. On the other hand, chemical treatment with excellent desliming effect has disadvantages, such as high operation costs and acid wastewater discharge, causing environmental problems, etc. Hence, the efficient utilization of NPQ resources on a large scale was delayed. Therefore, the removal of fine clay impurities from NPQ by an effective, environmentally friendly and affordable process is particularly critical for the large scale industrial utilization of NPQ.

Attrition scrubbing has been used as a physical decontamination treatment in many fields of mineral purification, especially in the removal of clay impurities from target mineral surfaces [14–16]. Centrifugal separation is one of the cost-effective solutions for the treatment of waste water, flotation slurries and chemical industry [17,18]. Hence, it is expected that the scrubbing-centrifugation technology could be an ideal method for the removal of clay impurities in NPQ.

In this paper, the feasibility of a combination technique of sieving, scrubbing and centrifugal separation on the removal of clay impurities in NPQ was investigated. In order to purify NPQ more efficiently, both NaOH and SHMP were added as regulators in the purification process. Furthermore, the possible mechanisms were also proposed by studying the heterocoagulation and homocoagulation between quartz and impurity particles according to the DLVO theory.

2. Materials and Methods

2.1. Materials

The raw ore of NPQ used in this study was obtained from JiangXi province, China. The main chemical composition was measured by ICP-AES (inductively-coupled plasma atomic emission spectrometry) and the result is presented in Table 1. As shown in Table 1, NPQ is mainly composed of SiO_2 (98.88%), and the main impurities are Al_2O_3 (0.48%) and Fe_2O_3 (0.19%); there are some other impurities like TiO_2, CaO, and MgO, etc. X-ray diffraction (XRD) was used to examine the purity of raw ore; the XRD pattern are presented in Figure 1. Figure 1 shows that the main peaks match well with the standard quartz peaks. However, no clear impurity peaks were observed, since the impurity content is too low, i.e., it exceeded the detection limit of XRD analysis.

Table 1. Chemical analysis of the dry NPQ (wt %).

Compositions	SiO_2	Al_2O_3	Fe_2O_3	TiO_2	CaO	MgO	K_2O	Na_2O	L.O.I. [a]
Content (wt %)	98.88	0.48	0.19	0.03	0.06	0.04	0.04	0.05	0.24

[a] Loss on ignition.

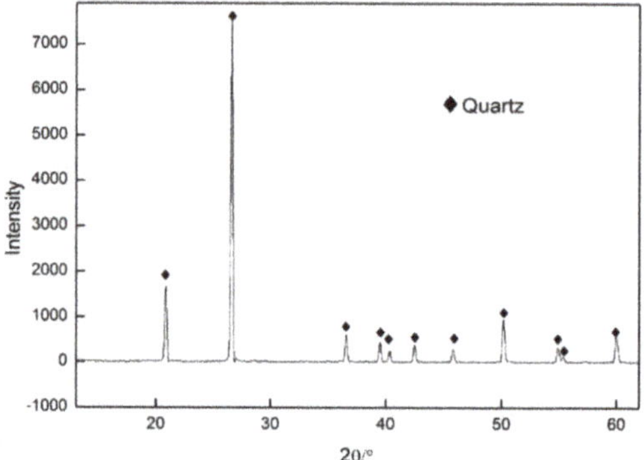

Figure 1. XRD pattern of NPQ.

All the reagents used in this study were chemically pure and purchased from Beijing Chemical Reagent Company (Beijing, China). Distilled water was used throughout the experiments.

2.2. Purification Methods

Highly concentrated (about 75%) pulp was prepared by adding 175 g water to 500 g raw ore, and then scrubbing tests were performed in vessels for about 25 min with the stirring rate of 1500 rpm. An appropriate amount of regulators (NaOH or SHMP) was pre-dissolved in 5 mL of water and then added to the scrubbing suspension at room temperature. After scrubbing, the pulp was stirred and dispersed for 5 min with extra 1750 mL water before being sieved using a 200 mesh standard screen (75 μm); the oversized product was tailing and labelled as "T1". The under-sized product was centrifuged at 315 rpm (RCF (relative centrifugal force) =10) for 1 min. The sediment of centrifugation was concentrated and labelled "Con", and the suspension of centrifugation was tailing and labelled "T2". All products were dried and weighed for analysis.

2.3. Equipment

The chemical compositions of the raw NPQ and purification products were measured using an ICP-OES (Agilent730, Agilent Technologies, Santa Clara, CA, USA). The crystalline phase identification of samples were undertaken by X-ray diffraction analysis D8 ADVANCE (Bruker, Billerica, MA, USA), equipped with a Cu-Kα radiation at a goniometer rate of $2\theta = 4°/min$ and the crystalline phase identification of trace impurities were detected by MLA. Microstructures were investigated by scanning electron microscopy (HT7700, Hitachi, Tokyo, Japan) at 10.0 KV, equipped with energy-dispersive microanalysis facilities (EDS). Zeta (ζ) potentials were measured using a Brookhaven zeta potential analyzer and three runs of measurement were conducted. The Mineral Liberation Analyser (MLA250, FEI, Hillsboro, OR, USA) was provided by Beijing General Research Institute of Mining & Metallurgy (BGRIMM, Beijing, China).

3. Results and Discussion

3.1. Impurity Analysis of NPQ

In order to analyze the distribution of impurities in different sizable fractions, NPQ were first sieved using a set of standard screens (100, 200 and 325 mesh), and then the part of −325 mesh was divided into sediment and suspension by centrifugation at 315 rpm for 1 min. The yields and contents of Fe_2O_3 and Al_2O_3 in different sizes of NPQ are summarized in Table 2. As displayed in Table 2, the yields were 1.96%, 4.51% and 5.44% for +100 mesh, −100~+200 mesh and micro-fine (centrifugal suspension) samples, respectively. The contents of Fe_2O_3 and Al_2O_3 in coarse grains are within the range of 0.132–0.301%and 0.157–0.862%, and in micro-fine the contents of Fe_2O_3 and

Al_2O_3 are 2.576% and 8.790%, respectively. On the other hand, the yields of $-200\sim+325$ mesh and the sediments of -325 mesh are 26.14% and 61.95% respectively with relatively lower content of impurities. It was concluded that the impurities of Fe_2O_3 and Al_2O_3 were mainly distributed in coarse grains (>200 mesh) and micro-fine grains.

Table 2. Yields and contents of Fe_2O_3 and Al_2O_3 in different sizable of NPQ.

Particle Size/Mesh		Yield (%)	Content (%)	
			Fe_2O_3	Al_2O_3
+100		1.96	0.301	0.862
$-100\sim+200$		4.51	0.132	0.157
$-200\sim+325$		26.14	0.040	0.031
-325	sediments	61.95	0.053	0.043
	suspensions	5.44	2.576	8.790

SEM equipped with EDS analysis was used to directly observe the microstructural characteristics and confirm the elements compositions of clay particles existing in NPQ. As shown in Figure 2, the dispersion of quartz was very even and the particle sizes were approximately in the range of 20–55 μm. Meanwhile, a large number of small particles (1–3 μm) were found adhering to the quartz surface. The EDS analysis showed that the main elements of particles on the quartz surface are Si, O, Al and Fe, indicating that the fine particles could be clay impurities.

Figure 2. (a) SEM image of NPQ, (b) SEM image of impurity adhering on quartz surface, (c) EDS of impurity.

Mineral liberation analysis (MLA) is a commercially available SEM-based image analysis system which has been widely applied in mineralogy and metallurgical processing [19,20]. Each mineral grain delineation inside the grain can be identified automatically with single X-ray analysis. The SEM-based image (BSE) signal in MLA has higher spatial resolution (0.1–0.2 μm) than X-ray of 2–5 μm. Hence, the mineralogy and the content of impurities in NPQ could not be determined by XRD because of the limitations of analytical accuracy, but could be determined by MLA. The representative results are displayed in Figures 3 and 4, and the distribution of Fe_2O_3 and Al_2O_3 according to mineralogy semi-quantitatively is summarized in Table 3. It was revealed that the presence of Fe_2O_3 comes mainly from hematite, and Al_2O_3 comes from orthoclase and anorthose. In addition, impurities such as mica were present in the sample.

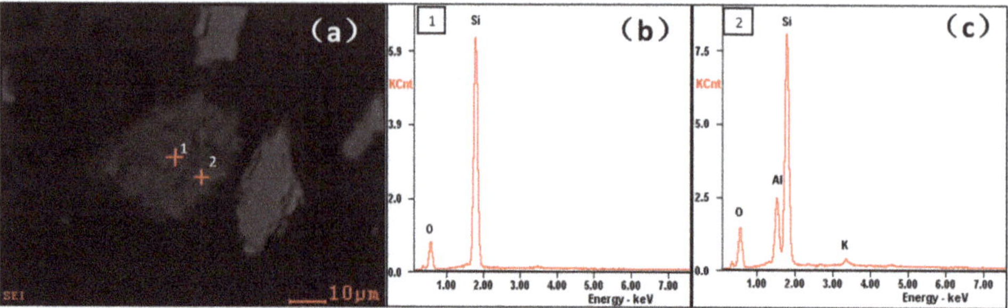

Figure 3. SEI image and spectrum of orthoclase in NPQ. (**a**) SEI image of quartz and orthoclase, (**b**) Energy spectrum of quartz, (**c**) Energy spectrum of orthoclase.

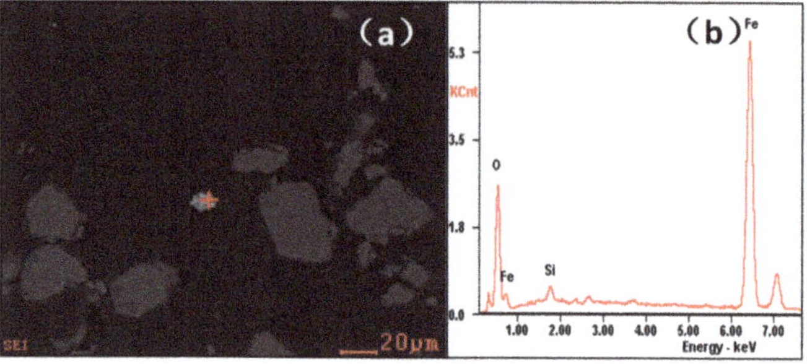

Figure 4. SEI image and spectrum of hematite in NPQ. (**a**) SEI image of hematite, (**b**) Energy spectrum of hematite.

Table 3. Distribution of Al_2O_3 and Fe_2O_3 at different minerals in NPQ detected by MLA (%).

Impurities	Hematite	Orthoclase	Anorthose	Mica	Total
Fe_2O_3	92.17	0	0	7.83	100
Al_2O_3	0	64.42	26.53	9.06	100

3.2. Effects of Scrubbing and Regulators on Impurities Removal

According to the character of impurities existing in the sample, a series of experiments were conducted as displayed in Table 4 to explore the effect of scrubbing and regulators on impurity removal. Firstly, experiments with and without scrubbing were performed. Subsequently, different regulators were added to the pulp when scrubbing. Test No. 1 was stirred 25 min with equivalent water and separated under the same condition as the test No. 2. The yields and element contents of Fe_2O_3 and Al_2O_3 in different products are summarized in Table 4.

The results displayed that the content of Fe_2O_3 and Al_2O_3 in concentrate without scrubbing decreased from 0.048% and 0.040% to 0.020% and 0.033% after scrubbing, respectively. Adding NaOH or SHMP when scrubbing could further decrease the content of Fe_2O_3 in concentrate to around 100 μg/g (0.01%) and the content of Al_2O_3 to around 200 μg/g (0.018–0.024%). The yield of both T2 and concentrate with scrubbing increased from 5.44% and 88.09% to 5.85% and 89.63%, and the yield of T1 decreased from 6.47% to 4.52%. Compared with scrubbing without regulators, adding regulators has little effect on the yields of products.

Table 4. Experiments conditions and results.

No.	Regulator	Dosage /(g·t^{-1})	Product	Yield (%)	Impurity Content (%) Fe_2O_3	Al_2O_3
1	/	/	T1	6.47	0.182	0.365
			T2	5.44	2.551	7.76
			Con	88.09	0.048	0.04
2	/	/	T1	4.52	0.173	0.311
			T2	5.85	2.857	7.51
			Con	89.63	0.020	0.033
3	NaOH	500	T1	4.33	0.141	0.179
			T2	6.03	2.948	7.602
			Con	89.64	0.010	0.018
4	SHMP	500	T1	4.38	0.148	0.169
			T2	6.02	2.963	7.521
			Con	89.60	0.010	0.024

The impurity distribution ratio (P) in different products under different conditions was calculated according Equation (1), which revealed the impurity change from a comprehensive view of content and yield, where Y_{Pro} and G_{Pro} represent the yield and the contents of Fe_2O_3 or Al_2O_3 in different products respectively, and C_{raw} represents the impurity content in raw ore.

$$P = \frac{Y_{Pro} \times G_{Pro}}{C_{raw}} \times 100\% \tag{1}$$

The results in Figure 5 suggest that P_{Fe2O3} and P_{Al2O3} of T1 and concentrate all decreased significantly after scrubbing, and thus, the P_{Fe2O3} and P_{Al2O3} of T2 both increased. It could be inferred that the physical impact and shearing actions occurring between contaminated particles and either the liquid phase or the walls and impellers stripped the clay impurities from quartz surface and crushed the coarse grains as micro-size fractions. The increase of P_{Fe2O3} and P_{Al2O3} of T2 indicated that more impurities were separated into T2 by centrifugation. The experimental results of test No. 3 and 4 showed that the concentrations of both P_{Fe2O3} and P_{Al2O3} further decreased, indicating that the separation efficiency of clay impurities from NPQ was improved by adding NaOH or SHMP.

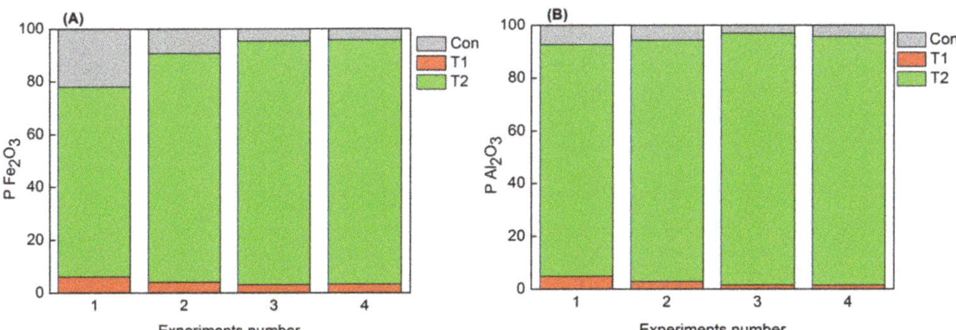

Figure 5. Distribution ratio of Fe_2O_3 (**A**) and Al_2O_3 (**B**) in T1, T2 and concentrate (Con) at different experiment conditions.

Some representative SEM images shown in Figure 6 were applied to identify the effect of scrubbing and centrifugation on the clay impurity removal of NPQ. As displayed in Figure 6 c,d, it is obvious that the fine clay particles adhering on quartz surface disappeared, and the surface was clearer and smoother after scrubbing compared with the samples which had not undergone scrubbing (Figure 6 a,b). It could be concluded that the impurity minerals were effectively stripped from the surface of quartz through scrubbing. According to Figure 6 e,f, it was demonstrated that the fine particles were separated into the T2 by centrifugation, which is the key factor affecting the removal efficiency of the clay impurities.

Figure 6. SEM images of concentrate and T2 in different conditions, (**a,b**) concentrate without scrubbing, (**c,d**) concentrate with scrubbing, (**e,f**) T2 of No. 2.

3.3. Effects of pH and Dosage of SHMP on Impurities Removal

To reveal the influence of pH on the removal efficiency of clay impurities, the impurities content of Fe_2O_3 and Al_2O_3 in concentrate were illustrated as a function of pH in Figure 7. The pH of solution was adjusted through adjusting NaOH amount, and the practical pH values were measured before centrifugation. The results showed that when the pH increased from 7.1 to 9.3, both of the contents of Fe_2O_3 and Al_2O_3 decreased significantly, but the increase of pH from 9.3 to 11.7 only shows slight influence on impurities removal.

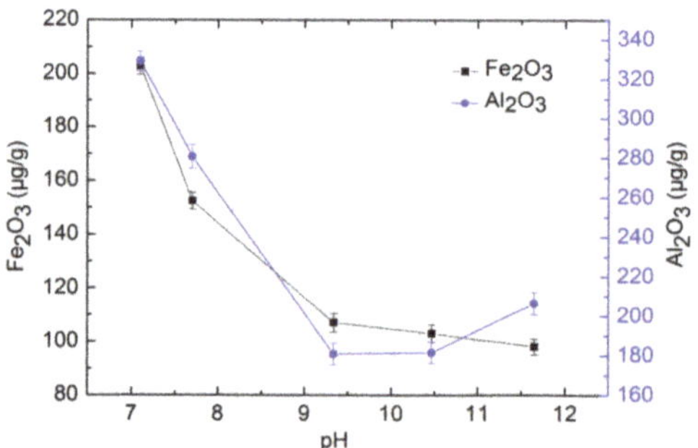

Figure 7. Effect of pH on the removal of impurities in concentrate.

The influence of SHMP on impurities removal was studied for dosages of 100, 500, 1000, 3000 and 5000 g/t at pH = 7 and the results are given in Figure 8. As expected, the dosage of SHMP affects both of the content of Fe_2O_3 and Al_2O_3 in concentrate. The removal efficiency of impurities was improved with the dosage of SHMP increases from 100 to 1000 g/t, and an optimal effect can be obtained at 1000 g/t. However, with increasing the dosage of SHMP continuously, the content of both Fe_2O_3 and Al_2O_3 in the concentrate presented an increasing trend, which indicated that an excess of SHMP can inhibit the removal efficiency of impurities.

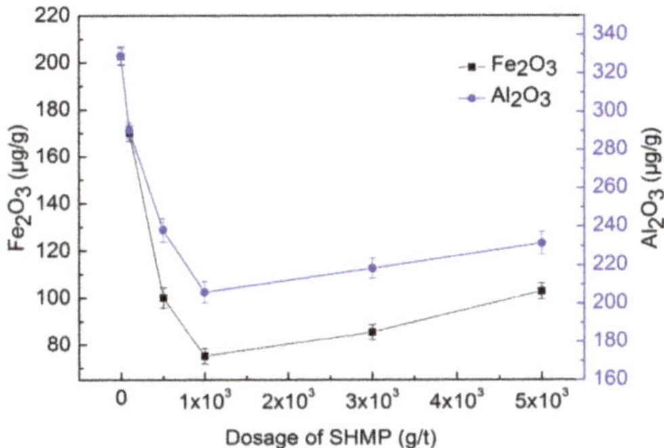

Figure 8. Effect of SHMP on the removal efficiency of impurities in concentrate.

4. Purification Mechanism

It is obvious that scrubbing enhanced the liberation of minerals, and the separation efficiency was dominated by centrifugation. Since previous experiments revealed that adding NaOH or SHMP could improve the removal efficiency of clay impurities, the mechanism was analyzed based on the influence of NaOH or SHMP on the interaction between quartz and clay impurity particles including hematite and orthoclase. Classical DLVO theory was used to study the stability of slurry and mineral particles by many researchers [21,22]. Based on the DLVO theory, the interactions between particles in aqueous solution include the van der Waals attraction force and the electrostatic double layer repulsive force. The coagulation and dispersion behaviors of mineral particles are mainly determined by the total interaction energy (V_T), which is the sum of the London-van der Waals dispersion interaction energy (V_A) and electrostatic repulsive energy (V_R), i.e., Equation (2), and the interaction energy V_A and V_R between two spheres can be calculated by Equations (3) and (4) [23].

$$V_T = V_R + V_A \tag{2}$$

$$V_R = \frac{-A_{132}}{6D} \frac{R_1 R_2}{R_1 + R_2} \tag{3}$$

$$V_A = \frac{\pi\varepsilon_0\epsilon R_1 R_2}{R_1 + R_2}\left\{ 4\varphi_1\varphi_2 ln\frac{1 + e^{-kD}}{1 - e^{-kD}} + \left\{ \varphi_1^2 + \varphi_2^2 \right\} ln\left\{ 1 - e^{-2kD} \right\} \right\} \tag{4}$$

where R_1 and R_2 are the radius of two spheres, D is the separation distance of two spheres. A_{132} is the effective Hamaker constant of substances 1 and 2 interacting in medium 3; it can be calculated by Equation (5), using the Lifshitz approach based on quantum physics [24]. φ_1 and φ_2 are the surface potentials of substance 1 and 2, which are substituted by the measured ζ potential value, κ^{-1} is the double layer thickness (9.6 nm in 1 mM KCl). ε is the dielectric constant of the aqueous medium (78.4 $C^2 \cdot m \cdot J^{-1}$), ε_0 is the dielectric constant of vacuum (8.854 $\times$ 10^{-12} $C^2 \cdot m \cdot J^{-1}$), the dielectric constants of quartz, hematite and orthoclase are 4.3, 6.9 and 81, respectively [25,26].

$$A_{total} = A_{v=0} + A_{v>0} \approx \frac{3}{4}kT\left(\frac{\varepsilon_1-\varepsilon_3}{\varepsilon_1+\varepsilon_3}\right)\left(\frac{\varepsilon_2-\varepsilon_3}{\varepsilon_2+\varepsilon_3}\right) + \frac{3h\nu_e}{8\sqrt{2}} \frac{\left(n_1^2-n_3^2\right)\left(n_2^2-n_3^2\right)}{\sqrt{\left(n_1^2+n_3^2\right)}\sqrt{\left(n_2^2+n_3^2\right)}\left\{\sqrt{\left\{n_1^2+n_3^2\right\}}+\sqrt{\left\{n_2^2+n_3^2\right\}}\right\}} \tag{5}$$

where k_B is the Boltzmann constant (1.381 $\times$ 10^{-23} $J \cdot K^{-1}$), T is the room temperature (293.15 K), h is the Planck constant (6.626 $\times$ 10^{-34} $J \cdot s$), ν_e is the main electronic absorption frequency in the UV typically around 3 $\times$ 10^{15} s^{-1} and n is the refractive index.

4.1. Effects of pH on the Zeta Potential

In this study, the standard minerals of hematite and orthoclase were selected as the main objects of impurity to analyze the processing mechanism. The standard minerals were obtained from commercial vendors. The quartz used in this paper was the concentrate of test No. 3, which was washed several times with distilled water. All the samples were ground and sieved by a 400 mesh for the test.

The ζ potentials of quartz, hematite and orthoclase were measured at different pHs, from neutral to alkaline, with a fixed ionic strength of 1.0 mM KNO_3. As shown in Figure 9, the ζ potentials of both hematite and orthoclase decreased as the pH increased, which were in agreement with the results reported by other researchers [27,28]. Unlike the characteristic of ζ potentials of hematite and orthoclase, one interesting feature has been observed from the plot of ζ potential of quartz. Initially, the ζ potential of quartz decreased as the pH increased from 7 to 9. However, when the NaOH was added to the slurry to increase the pH above 9, this potential of quartz gradually increased; this phenomenon also has been detected by other researchers [29]. The possible reasons for the charge changing of quartz according to the reference could be explained as follows: the metal ions in solution and the clay particles coated on the quartz surface, or the solubility of quartz rises sharply with increasing the solution pH, which could compressed the electric double layers of minerals, and thus, decreased the negative charge potentials [30–32].

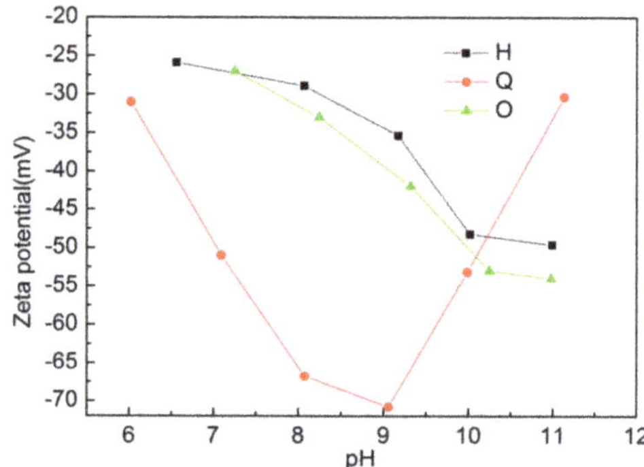

Figure 9. The ζ potential of quartz (Q), hematite (H) and orthoclase (O) as a function of pH at constant ionic strength of 1.0 mM KNO_3.

4.2. Heterocoagulation between Quartz and Impurities

The interaction energy between quartz and impurity particles was calculated based on the DLVO theory firstly. It is important to emphasize that the impurity particles (radius 3 μm) and NPQ (radius 40 μm) were assumed as two ideal spheres and the measured ζ potential value as the surface potential. Figure 10 showed the interaction energy of V_R, V_A and V_T respectively between quartz and hematite (Q–H), quartz and orthoclase (Q–O) at different pHs. When the separation distance was above 3 nm, both of the V_T of Q–H and Q–O were dominated by V_R, and thus, the values were positive as pHs varied from 7 to 11, which suggested that the interaction between quartz and impurities was primarily repulsive in nature. It is also indicated that the V_T change to negative with the V_A tend to negative infinitely when interaction distance being much closer (less than 1 nm), which means that the interaction force was attractive when the impurity particles adhering on quartz surface. The results also illustrated the importance of scrubbing, which provided sufficient energy to overcome the strong V_A between quartz and impurities adhering on quartz surface.

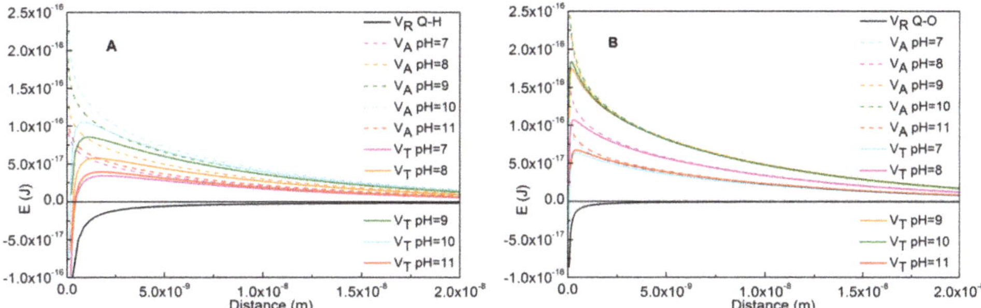

Figure 10. Relationship between interaction energy of impurities with quartz and particles distance at different pH; (**A**) is the relationship of hematite with quartz; (**B**) is the relationship of orthoclase with quartz.

Although the V_T between quartz and impurities always remained positive, the experimental results showed that the removal efficiency of impurities has a strong correlation with the changing of V_T. As can be seen, the V_T was increased gradually with the pH increase from 7 to 9, which is consistent with the changing trend of removal efficiency of impurities according to the previous experiments results. The reason could be explained that the V_T was calculated based on the assumption that the averaging particle size of impurity was 3 μm, but the practical particle size should be normally distributed. According to Equation (2), V_T is roughly proportional to the radius of impurity particle R_1 when the radius of quartz R_2 and the ζ potential of minerals are fixed. When the radius of impurity tends to 0, V_T also tends to zero, which means that there are always a certain proportion of fine impurity particles which could be attracted on the quartz surface easily, and the proportion of particles attached on quartz would decrease with the increase of V_T.

At pH 10, even though the ζ potential of quartz displayed a reverse increase, V_T keeps increasing as the ζ potentials of both hematite and orthoclase decreased, and the effect of pH from 9 to 10 on impurities removal is not notable. When pH was up to 11, V_T reduced significantly, and consequently, a worse impurity removal effect was observed. All the results indicated that increasing V_T by adjusting pH could improve the impurities removal. However, the strong alkaline environment was not beneficial for impurity removal, as the reverse increasing of the ζ potential of quartz.

4.3. Coagulation of Fine Impurity Particles

Apart from the heterocoagulation between quartz and impurity particles, the interaction among fine impurity particles was also calculated to analyze the coagulation, which could also influence impurity removal. Figure 11 indicates the interaction energy between hematite and hematite (H–H), orthoclase and orthoclase (O–O), orthoclase and hematite (O–H) under natural condition (pH = 7). The results showed that the V_T of O–O and O–H were both positive, while the V_T of H–H was negative over the whole separation distance, which indicated that the coagulation of fine particles occurred mainly by the formation of homocoagulation between H–H in the slurry.

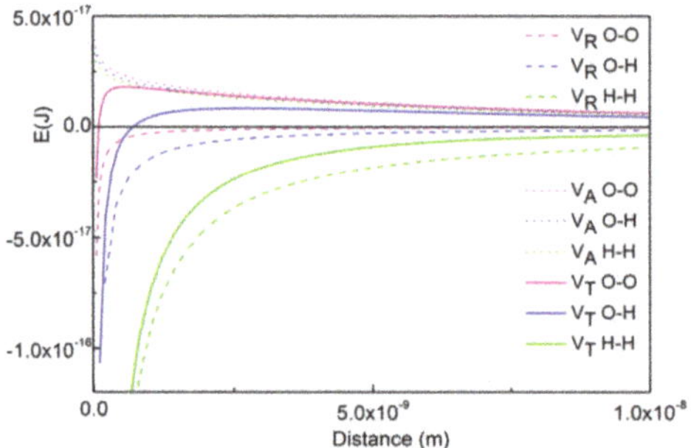

Figure 11. The relationship between interaction energy of fine impurities and particles distances at pH 7.

V_T of H–H at different pHs is shown in Figure 12. As can be seen, V_T was negative at pH 7 and 8. When the pH increased to 9, V_T was close to zero and then changed to positive at pH 10 and 11. The results indicated that the homocoagulation of H–H occurred more easily when the pH was 7 and 8, and then became dispersive at pH > 9. The conclusion was reinforced by the observation about the stability of slurry of T2, as shown in Figure 13, which showed the settlement of T2 after centrifugation for 24 h at different pH values. As can be seen, when the pH was lower than 9, the fine particles in the slurry were coagulated, and consequently, they settled more easily. On the other hand, when the pH was over 9, the slurry became more stable, which was favorable to the removal of these impurities.

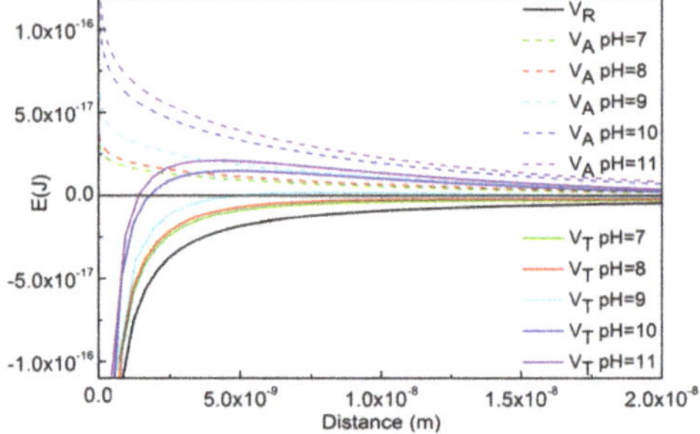

Figure 12. The relationship between interaction energy of H–H and particles distance at different pHs.

Figure 13. Stability of slurry of T2 at different pHs.

4.4. Effects of SHMP on the Zeta Potential and Desliming

SHMP is a kind of long-chain inorganic salt that has been used as a dispersant for clay mineral separation, which could increase the negative zeta potential of particles, and thus improve the dispersion of slurry [33,34]. The ζ potentials of quartz, hematite and orthoclase were measured at the optimized proportion of SHMP (1×10^3 g·t^{-1}), according to the previous experiments. As can be seen from Table 5, the addition of SHMP exhibited a remarkable effect on decreasing the ζ potential of all three minerals. As stated in the previous analysis, the ζ potentials of minerals were negative enough to lead to a good dispersion of slurry, which is agreement with the impurity removal results, as shown in Figure 8.

Table 5. ζ potential of quartz, hematite and orthoclase at different dosage of SHMP at pH 7.

ζ	Dosage of SHMP /g·t^{-1}		
	0	1000	5000
Quartz	−51.31	−56.03	−38.57
Hematite	−27.01	−61.83	−51.23
Orthoclase	−25.56	−75.07	−67.57

Moreover, ζ potentials of quartz, hematite and orthoclase at the dosage of SHMP being 5×10^3 g·t^{-1} were measured to explain the destabilization of impurity removal when excessive SHMP were added. When the dosage of SHMP increased from 1×10^3 g·t^{-1} to 5×10^3 g·t^{-1}, as can be seen, ζ potentials of all three minerals present an inverse increase (absolute value decrease). It should also be noted that the electrical double layers of minerals were significantly compressed at high ionic strengths [35]. All the results indicate that an appropriate dosage of SHMP could improve the removal efficiency of clay impurities in NPQ, but an excessive dosage of SHMP could have a negative impact on the impurity removal, i.e., the same as adjusting pH.

5. Conclusions

In summary, scrubbing combining with centrifugation was proven to be an effective method for the desliming purification of NPQ. The removal efficiency of impurities reached an optimal value at pH 9.3, or when the dosage of SHMP reached 1×10^3 g/t. The main impurities of Fe_2O_3 and Al_2O_3 in the concentrate decreased to less than 0.01% and 0.02%, respectively. It was concluded that the scrubbing process played an important role in the enhancement of impurities liberation. Then, the addition of regulators such as NaOH or SHMP could effectively increase the negative ζ potential of minerals, and thus increase the V_T, which could effectively decrease the possibility of the heterocoagulation between quartz and impurities, as well as the homocoagulation among hematite particles, consequently improving the removal efficiency of impurities. In conclusion, this technique could be an environment-friendly and highly efficient method for the purification of NPQ.

Author Contributions: Conceptualization, Z.S. and S.Z.; Data Curation, X.D.; Investigation, X.D., C.L. and D.H.; Project Administration, Z.S. and S.Z.; Software, X.D.; Supervision, Z.S. and S.Z.; Writing—Original Draft, X.D.; Writing—Review and Editing, Z.S. and S.Z.

Funding: This research was funded by the Young Elite Scientists Sponsorship Program by CAST (2017QNRC001), the Fundamental Research Funds for the Central Universities (2010YH10 and 2015QH01)

Conflicts of Interest: The authors declare no conflict of interest.

References

1. Santos, M.F.M.D.; Fujiwara, E.; Schenkel, E.A.; Enzweiler, J.; Suzuki, C.K. Quartz sand resources in the Santa Maria Eterna formation, Bahia, Brazil: A geochemical and morphological study. *J. South Am. Earth Sci.* **2015**, *62*, 176–185. [CrossRef]
2. Huang, H.; Li, J.; Li, X.; Zhang, Z. Iron removal from extremely fine quartz and its kinetics. *Sep. Purif. Technol.* **2013**, *108*, 45–50. [CrossRef]
3. Prakash, S.; Das, B.; Mohanty, J.K.; Venugopal, R. The recovery of fine iron minerals from quartz and corundum mixtures using selective magnetic coating. *Int. J. Miner. Process.* **1999**, *57*, 87–103. [CrossRef]

4.	Zhong, L.L.; Lei, S.M.; Wang, E.W.; Pei, Z.Y.; Li, L.; Yang, Y.Y. Research on Removal Impurities from Vein Quartz Sand with Complexing Agents. *Appl. Mech. Mater.* **2013**, *454*, 194–199. [CrossRef]

5.	Andrews, P.R.A.; Collings, R.K. Canadian silica resources for glass and foundry sand production: Processing studies at CANMET. *Int. J. Miner. Process.* **1989**, *25*, 311–317. [CrossRef]

6.	Tuncuk, A.; Akcil, A. Iron removal in production of purified quartz by hydrometallurgical process. *Int. J. Miner. Process.* **2016**, *153*, 44–50. [CrossRef]

7.	Zhang, Z.; Li, J.; Li, X.; Huang, H.; Zhou, L.; Xiong, T. High efficiency iron removal from quartz sand using phosphoric acid. *Int. J. Miner. Process* **2012**, *114–117*, 30–34. [CrossRef]

8.	Hou, Y.; Liu, P.; Hou, Q.; Duan, H.; Xie, Y. Study on removal fluid inclusions in quartz sand by microwave explosion. *Nanosci. Nanotechnol. Lett.* **2017**, *9*, 151–154. [CrossRef]

9.	Štyriakov, A.I.; Štyriak, I.; Kraus, I.; Uhlík, P.; Madejov, A.J.; Orolínov, A.Z. Bioleaching of clays and iron oxide coatings from quartz sands. *Appl. Clay Sci.* **2012**, *61*, 1–7. [CrossRef]

10.	He, M.S.; Yao, Q.; Yi, Y.Q. Analysis of the development and utilization and the supply and demand situation on Natural powder quartz at Jiang Xi province. *China Nonmet. Miner. Ind.* **2003**, *6*, 52–55. (In Chinese)

11.	Richard, B.; John, M.M. Geology of Microcrcrystalline Silica (Tripoli) Deposits, Southernmost Illinois. Illinois State Geological Survey. 1994; pp. 4–8. Available online: https://core.ac.uk/download/pdf/17355415.pdf (accessed on 27 January 2019).

12.	Tan, J.; Zhou, H.; Wang, M.; Zhu, B.Q.; Gao, X.Q.; Zheng, C.H. Study of the purification and whitening of Natural powder quartz. *Acta Mineral. Sin.* **2014**, *34*, 7–12. (In Chinese)

13.	Zhang, Y.P.; Huang, K.L.; Liu, S.Q. Separation of clinochlore from powder quartz by reverse flotation and its mechanism. *J. Cent. South Univ. Technol. (Sci. Technol.)* **2007**, *2*, 285–290. (In Chinese)

14.	Bayley, R.W.; Biggs, C.A. Characterisation of an attrition scrubber for the removal of high molecular weight contaminants in sand. *Chem. Eng. J.* **2005**, *111*, 71–79. [CrossRef]

15.	Stražišar, J.; Sešelj, A. Attrition as a process of comminution and separation. *Powder Technol.* **1999**, *105*, 205–209. [CrossRef]

16.	Sun, Z.; Yang, X.; Zhang, G.; Zheng, S.; Frost, R.L. A novel method for purification of low grade diatomite powders in centrifugal fields. *Int. J. Miner. Process.* **2013**, *125*, 18–26. [CrossRef]

17.	Batalović, V. Centrifugal separator, the new technical solution, application in mineral processing. *Int. J. Miner. Process.* **2011**, *100*, 86–95. [CrossRef]

18.	Kang, S.; Zhao, Y.; Wang, W.; Zhang, T.; Chen, T.; Yi, H.; Rao, F.; Song, S. Removal of methylene blue from water with montmorillonite nanosheets/chitosan hydrogels as adsorbent. *Appl. Surf. Sci.* **2018**, *448*, 203–211. [CrossRef]

19.	Petruk, W. Automatic Image Analysis for Mineral Beneficiation. *JOM* **1988**, *40*, 29–31. [CrossRef]

20.	Fandrich, R.; Gu, Y.; Burrows, D.; Moeller, K. Modern SEM-based mineral liberation analysis. *Int. J. Min. Process.* **2007**, *84*, 310–320. [CrossRef]

21.	Houta, N.; Lecomte-Nana, G.L.; Tessier-Doyen, N.; Peyratout, C. Dispersion of phyllosilicates in aqueous suspensions: Role of the nature and amount of surfactant. *J. Colloid Interface Sci.* **2014**, *425*, 67–74. [CrossRef] [PubMed]

22.	Yu, Y.; Ma, L.; Xu, H. DLVO theoretical analyses between montmorillonite and fine coal under different pH and divalent cations. *Powder Technol.* **2018**, *330*, 147–151. [CrossRef]

23.	Israelachvili, J.N. *Intermolecular and Surface Forces*, 3rd ed.; Academic Press: New York, NY, USA, 1985; p. 262.

24.	Berg, J.C. *An Introduction to Interfaces & Colloids: The Bridge to Nanoscience*; World Scientific: Singapore, 2010.

25.	Robinson, D. Measurement of the solid dielectric permittivity of clay minerals and granular samples using a time domain reflectometry immersion method. *Vadose Zone J.* **2004**, *3*, 705–713. [CrossRef]

26.	Marland, S.; Merchant, A.; Rowson, N. Dielectric properties of coal. *Fuel* **2001**, *80*, 1839–1849. [CrossRef]

27.	Kaya, A.; Yukselen, Y. Zeta potential of clay minerals and quartz contaminated by heavy metal. *Can. Geotech. J.* **2005**, *42*, 1280–1289. [CrossRef]

28.	Yukselen-Aksoy, Y.; Kaya, A. A study of factors affecting on the zeta potential of kaolinite and quartz powder. *Environ. Earth Sci.* **2011**, *62*, 697–705. [CrossRef]

29.	Rohem Peçanha, E.; da Fonseca de Albuquerque, M.D.; Antoun Simão, R.; de Salles Leal Filho, L.; de Mello Monte, M.B. Interaction forces between colloidal starch and quartz and hematite particles in mineral flotation. *Colloid Surf. A Physicochem. Eng. Asp.* **2019**, *562*, 79–85. [CrossRef]

30. James, R.O.; Healy, T.W. Adsorption of hydrolyzable metal ions at the oxide—Water interface. II. Charge reversal of SiO_2 and TiO_2 colloids by adsorbed Co(II), La(III), and Th(IV) as model systems. *J. Colloid Interface Sci.* **1972**, *40*, 53–64. [CrossRef]

31. Li, Z.-Y.; Xu, R.-K.; Li, J.-Y.; Hong, Z.-N. Effect of clay colloids on the zeta potential of Fe/Al oxide-coated quartz: A streaming potential study. *J. Soils Sediments* **2016**, *16*, 2676–2686. [CrossRef]

32. Rodríguez, K.; Araujo, M. Temperature and pressure effects on zeta potential values of reservoir minerals. *J. Colloid Interface Sci.* **2006**, *300*, 788–794. [CrossRef] [PubMed]

33. Espinozaortega, O. Role of Sodium Hexametaphosphate in the Flotation of Acanthite Fines from Finely Disseminated Ores. *Sep. Sci. Technol.* **2009**, *44*, 2971–2982.

34. Zhang, T.; Vandeperre, L.J.; Cheeseman, C.R. Formation of magnesium silicate hydrate (M-S-H) cement pastes using sodium hexametaphosphate. *Cem. Concr. Res.* **2014**, *65*, 8–14. [CrossRef]

35. Tiller, C.L.; O'Melia, C.R. Natural organic matter and colloidal stablility: Models and measurements. *Colloid Surf. A Physicochem. Eng. Asp.* **1993**, *73*, 89–102. [CrossRef]

Article

The Process of the Intensification of Coal Fly Ash Flotation Using a Stirred Tank

Lu Yang [1,2]**, Zhenna Zhu** [1,2]**, Xin Qi** [1,2]**, Xiaokang Yan** [2] **and Haijun Zhang** [1,]*

[1] Chinese National Engineering Research Center for Coal Preparation and Purification,
     China University of Mining and Technology, Xuzhou 221116, China; ylpass@163.com (L.Y.);
     zznmandy@163.com (Z.Z.); hgkjqx@163.com (X.Q.)
[2] School of Chemical Engineering and Technology, China University of Mining and Technology,
     Xuzhou 221116, China; xk-yan@cumt.edu.cn
*   Correspondence: zhjcumt@cumt.edu.cn; Tel.: +86-516-83885878

Received: 14 November 2018; Accepted: 14 December 2018; Published: 17 December 2018

Abstract: Pulp preconditioning using a stirred tank as a pretreatment process is vital to the flotation system, which can be used to improve the flotation efficiency of mineral particles. The kinetic energy that is dissipated in the stirred tank could strengthen the interaction process between mineral particles and flotation reagents to improve the flotation efficiency in the presence of the preconditioning. In this paper, the effect of the conditioning speed on the coal fly ash flotation was investigated using numerical simulations and conditioning-flotation tests. The large eddy simulation coupled with the Smagorinsky-Lilly subgrid model was employed to simulate the turbulence flow field in the stirred tank, which was equipped with a six blade Rushton turbine. The impeller rotation was modelled using the sliding mesh. The simulation results showed that the large eddy simulation (LES) well matched the previous experimental data. The turbulence characteristics, such as the mean velocity, turbulent kinetic energy, power consumption and instantaneous structures of trailing vortices were analysed in detail. The turbulent length scale (η) decreased as the rotation speed increased, and the minimum value of η was almost unchanged when the rotation speed was more than 1200 rpm. The conditioning-flotation tests of coal fly ash were conducted using different conditioning speeds. The results showed that the removal of unburned carbon was greatly improved due to the strengthened turbulence in the stirred tank, and the optimal results were obtained with an LOI of 3.32%, a yield of 78.69% and an RUC of 80.89% when the conditioning speed was 1200 rpm.

Keywords: intensification; coal fly ash; flotation; preconditioning; stirred tank

---

1. Introduction

Coal is the main energy source in China [1]. However, the direct combustion of coal in thermal power plants has released many pollutants into the environment [2]. Coal fly ash (FA) is one of the main by-products of pulverized coal combustion [3]. The unburned carbon (UC) content in FA, which is evaluated using conventional loss on ignition (LOI), might be an obstacle for the utilization of FA in a variety of industries, especially for its use as a raw material in cement and concrete [4]. The recommended LOIs of the Class 1 and 2 fly ash types used in concrete should be less than 5% and 8%, respectively, according to the GB/T 1596-2017 [5]. The UC collected from FA could be used as activated carbon materials, graphite-like materials, etc. [6,7]. Froth flotation is still the main method for the separation of unburned carbon from FA [8,9].

Froth flotation is an important beneficiation method to realize the separation of valuable minerals and gangue minerals. The essence of flotation preconditioning is the flow-transfer-adsorption(reaction) process with multiple components, multiple scales and three phases, where the process factor determines the flotation efficiency and ability. Currently, the research on the physicochemical characteristics on the interface of mineral-reagents and mineral-bubbles has achieved good results [10,11], while the process intensification mechanism caused by the fluid strengthening lacks in-depth study. Pulp conditioning as a pretreatment before the flotation operation is of great importance to the flotation system [12,13]. Feng et al. studied the effect of conditioning methods on the flotation of nickel ore and found that high intensity conditioning could increase pentlandite flotation recovery significantly compared with the addition of Calgon [14]. Yu et al. investigated the effect of agitation on the interaction of coal and kaolinite during flotation and found that mild agitation (0~1200 rpm) could enhance the kaolinite-coating, which would lead to lower combustible matter recovery, while high intensity agitation (1200–2000 rpm) mitigated the kaolinite-coating [15]. Tabosa and Rubio investigated copper sulphide flotation assisted by high intensity conditioning and the results showed that the high intensity conditioning, as a pre-flotation stage, could increase both the copper grade and recovery [16].

Stirred tanks always play an important role in flotation preconditioning [17]. The flow structure in stirred tanks is three-dimensional, complex and has high shear. The computational fluid dynamics (CFD) simulation is used to promote the study of the fundamental phenomena, which could provide the flow field information in the stirred tank [18]. As we know, the large eddy simulation (LES) could obtain the precision flow field information, and its computational consumption is between the direct numerical simulation (DNS) and the Reynolds-averaged Navier-Stokes equations (RANS) [19]. Large eddy simulation was first applied in the simulation of stirred tanks by Eggles and was proven to be a good method to study turbulent flows [20]. The basic idea of large eddy simulations is filtering. The turbulent movement could be decomposed into the resolved scales (which include the larger scale fluctuations) and the unresolved scales (which include all the smaller scale fluctations) through filtration. The turbulent movement of the resolved scale could be solved directly using the numerical calculation method. However, for the unresolved scale, the effects of the mass, momentum and energy transport on the larger scale movement are modelled using a subgrid-scale model, which can closely approximate the movement equation [21].

The turbulence flow in the stirred tank can be regarded as the formation of turbulent vortexes, which is the main driving force for fluid shear, dispersion and uniform mixing. The interaction process between particles and reagents is completed in the full turbulent flow field of the stirred tank since the turbulence has an important role in the interaction process. The turbulent fluctuations in the turbulence movement could be regarded as the simultaneous existence of energetic vortices with different length scales. The energetic vortices possess the vast majority of the fluid kinetic energy and pass the energy to the smaller vortices step by step. The energy input agitated by the impeller rotation is dissipated by the heat transfer due to the molecular viscosity of the fluid at the smallest vortices in the stirred tank. The smallest scale of energy dissipation of the turbulent vortices is also called the turbulent microscale (η). The kinetic energy that is dissipated in the stirred tank could strengthen the interaction process between the mineral particles and the flotation reagents to improve the flotation efficiency in the presence of preconditioning with a stirred tank.

It was validated that the hydrodynamic conditions had important influences on the breakup of droplets breakup [22], the particle agglomeration in crystallizers [23], the micromixing [24] and the passive scalar transport [25] in the processing performance. This paper focuses on the turbulent properties in the outflow generated by the Rushton turbine in a stirred tank and the conditioning-flotation tests of the flotation separation of unburned carbon from coal fly ash. The simulation results were compared with previous studies to validate the reliability of the large eddy simulation (LES). The trailing vortices and the turbulence length scale under different conditioning speeds were investigated. Then, conditioning-flotation tests were carried out to verify the optimal turbulence flow field with the purpose to improve the separation efficiency of unburned carbon from coal fly ash.

2. Methodology

2.1. Geometry Structure

The stirred tank was a standard cylindrical vessel of diameter T = 300 mm, which was stirred by a six blade Rushton turbine, was equipped with four equi-spaced baffles of width 1/12 T and had a liquid height of H = T, as shown in Figure 1. The impeller had a diameter of D = T/3 and a clearance of C = T/3. The working fluid was water with a density of 1000 kg/m^3 and a kinematics viscosity of 1.006×10^{-6} m^2/s. The impeller rotational speeds were 300, 600, 900, 1200, and 1500 rpm, the Reynolds number ($R_e = ND^2/v$, v is the fluid kinematics viscosity and N is the impeller rotational speed) was 5.0×10^4 ~2.5×10^5, and the impeller tip speed was $U_{tip} = \pi DN$.

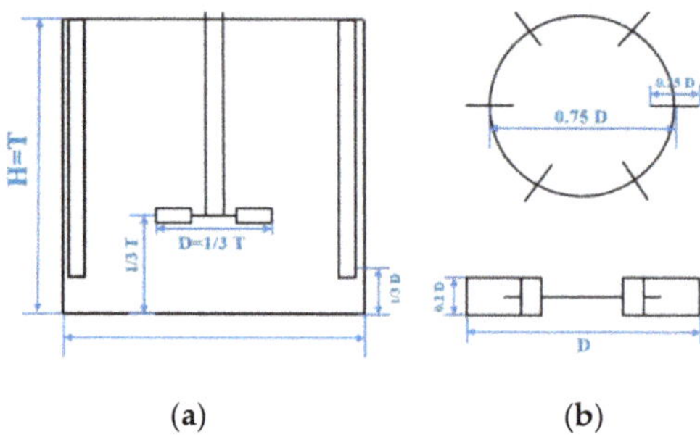

Figure 1. View of (**a**) the stirred tank and (**b**) the Rushton turbine.

2.2. Governing Equations

The LES can resolve the larger scale movements and model the smaller scale movements by employing a spatial filter for the governance of the Navier-Stokes equations. The governing equations of LES are as follows:

$$\frac{\partial \overline{u}_i}{\partial x_i} = 0 \tag{1}$$

$$\frac{\partial(\overline{u}_i)}{\partial t} + \overline{u}_i \frac{\partial(\overline{u}_j)}{\partial x_i} + \frac{\partial \overline{P}}{\partial x_i} = v \frac{\partial^2 \overline{u}_i}{\partial x_j^2} - \frac{\partial \tau_{ij}}{\partial x_j} \tag{2}$$

$$\tau_{ij} = \overline{u_i u_j} - \overline{u}_i \overline{u}_j \tag{3}$$

where τ_{ij} is the sub-grid scale stress tensor, which reflects the effect of the unresolved scale on the resolved scale [26]. In this study, the effect of the sub-grid scale on the larger scale was assessed using the Smagorinsky-Lilly model, and the eddy viscosity is modelled as $v_t = L_s^2|\overline{s}|$, where $L_s = \min(Kd, C_S V^{1/3})$, K is the mixing length for the subgrid scale, d is the distance to the closest wall, V is the volume of the computational cells, C_S is the Smagorinsky constant of 0.1 in this work [27–29], and $|\overline{s}| = \sqrt{2\overline{s}_{ij}\overline{s}_{ij}}$, $\overline{s}_{ij}$ is the rate of the strain tensor for the resolved scale.

2.3. Numerical Methods

The numerical simulations were conducted using the FLUENT software. A fine mesh with a high quality and a small time step should be used in the LES. The computational meshes were divided into two parts: a rotation cylindrical volume that included the impeller, and a stationary part that contained the rest of the tank. The computational meshes with homogeneous hexahedral structure were generated by the ICEM. A mesh independence test was carried out at the rotation speed of 300 rpm. The volume-averaged turbulence dissipation rate was chosen as the principle criterion of the mesh quality [30] and the results are shown in Figure 2. The figure shows that the volume-averaged turbulence dissipation rate had a slight change when the mesh quantity was larger than 0.86 million, and so the mesh quantity of 0.86 million was chosen for the numerical calculation at 300 rpm. On this basis, considering the computational demand, the mesh quantity, mainly in the vicinity of the impeller, was increased to approximately 1.20 million, 1.61 million, 2.21 million and 3.16 million for the numerical calculations at 600 rpm, 900 rpm, 1200 rpm and 1500 rpm, respectively. The mesh schematics at the rotation speed of 300 rpm are shown in Figure 3.

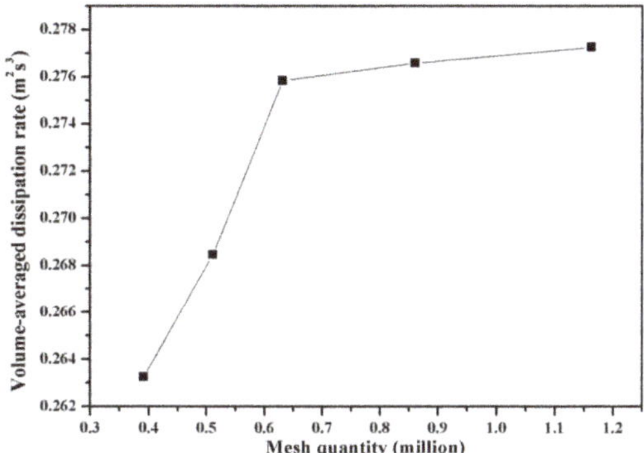

Figure 2. Volume-averaged turbulence dissipation rate for different mesh quantities.

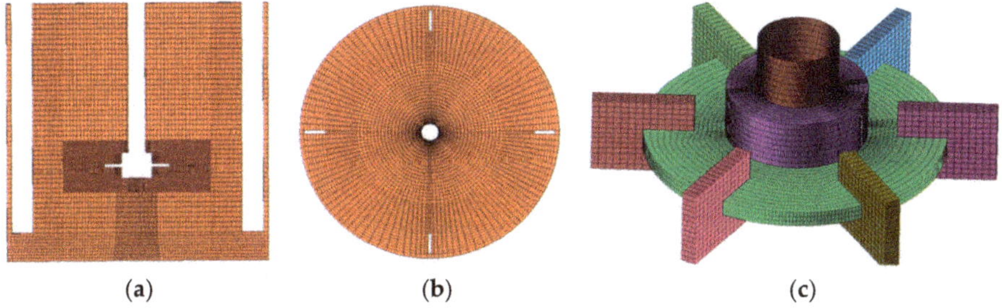

(a)   (b)   (c)

Figure 3. View of the hexahedral mesh. (**a**) vertical section; (**b**) horizontal section; (**c**) impeller surface.

The LES with the Smagorinsky-Lilly subgrid model was employed to simulate the flow field in the stirred tank. Sliding mesh (SM) was used to deal with the movement of the impeller. The complete stabilized flow field, which was obtained from the k-epsilon simulation, was used as the initial values of the LES. The second order and the bounded central differences were chosen as the spatial discretization of the pressure and momentum, respectively. The time steps used in the simulations were set at 0.001 for 300 rpm, 0.001 for 600 rpm, 0.0005 for 900 rpm, 0.0001 for 1200 rpm, and 0.0001 for 1500 rpm, which take into account the computational consumption. The total simulation time was 10 s at the agitation rates of 300 rpm and 600 rpm and 20 revolutions of the impeller for the other higher agitation rates for getting the steady flow field. Then, the flow data were collected and postprocessed. Most results stated here were the data for 300 rpm, except for special instructions.

2.4. The Post-Processing of the LES Results

Since LES solves an instantaneous velocity field, post-processing was performed to extract the mean velocity, the periodic velocity fluctuations and the turbulent velocity fluctuations according to their statistical averages using the MATLAB software. The velocity fluctuations in a turbulently stirred tank are partly periodic (or organized) and partly random (turbulent). The local instantaneous velocity u_i should be decomposed into the mean, the periodic and resolved the part of the fluctuations as follows:

$$u_i = \overline{U_i} + \tilde{u}_i + u_i' \tag{4}$$

As a result, the kinetic energy can be divided into the periodic and the turbulent kinetic energy. The total kinetic energy, k_{tot}, in the velocity fluctuations is as follows:

$$k_{tot} = k_P + k_T \tag{5}$$

where k_P and k_T are the periodic kinetic energy and turbulent kinetic energy with respect to the total turbulent kinetic energy k_{tot}, respectively. The turbulent kinetic energy is calculated as follows:

$$k_T = \frac{1}{2} \sum_{i=1}^{3} \langle u_i' u_i' \rangle \tag{6}$$

The dissipation rate of the turbulent kinetic energy can be calculated from the measurements of the turbulent velocity gradients in three directions as follows:

$$\varepsilon = v S_{ij} S_{ij} = \frac{v}{2} \left(\frac{\partial u_i}{\partial x_j} + \frac{\partial u_j}{\partial x_x} \right)^2 \tag{7}$$

3. Numerical Simulation Results and Discussion

3.1. Mean Velocity

The mean velocity of the LES results in a radial position of r/R = 1.07 in the impeller jet flow, which is normalized by the impeller tip velocity (U_{tip}), as shown in Figure 4. Figure 4 shows the comparison results between the simulations and previous experiments (LDV data from Wu, 1989 [31]; and PIV data from Escudié, 2001 [32]). 2z/W = 0 represented the central position of the impeller disc, and Ur, Ut and Uz represented the radial velocity, tangential velocity and axial velocity, respectively. The radial velocity of the LES agreed well with the previous experiments. For the simulation results and previous experiments, they have almost the same width of the flow entrainment area, and the peak location of the radial velocity was at 2z/W = 0. For the tangential velocity, the value of the peak in the simulation was slightly higher, while the location was in better agreement with the LDV results and slightly lower compared with the PIV results. In addition, a wider flow entrainment area was obtained compared with the experimental results. The axial velocity was evidently smaller than the radial velocity and the tangential velocity, and it was almost zero around the central position, which was because the fluid split into two streams in opposite directions. It could be seen that there was a difference between the velocity of the simulations and experiments at both ends of the jet flow area. The fine distinction of the mean velocity might be caused by the geometrical shape, experimental methods and operating conditions [33–35].

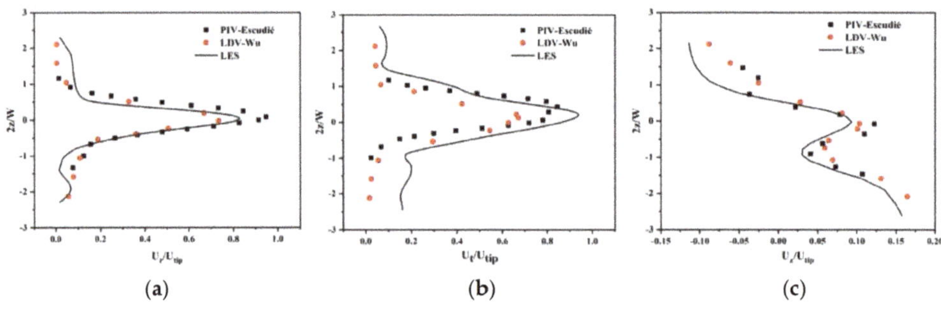

Figure 4. Mean velocity normalized by U_{tip}. (**a**) radial velocity; (**b**) tangential velocity; (**c**) axial velocity.

3.2. Turbulent Kinetic Energy

The dimensionless turbulent kinetic energy (k_T/U_{tip}^2) at the radial position of r/R = 1.07 was compared with the previous work (the PIV data from Escudié, 2001 [32]) to validate the LES under all rotation speeds, as shown in Figure 5. The turbulent kinetic energy was larger in the vicinity of the impeller, and it decreased as it moved away from the impeller. The upper impeller outflow was slightly larger than the lower impeller outflow in the simulations for all the rotation speeds. The width of the peaks of the LES simulations and PIV data were roughly aligned. It can be seen that a good agreement was found for the simulation results and PIV data. There was also a difference between the LES results and the previous experimental data, which may be due to the insufficient spatial resolution of the mesh size and the difference in the geometrical shapes [36,37].

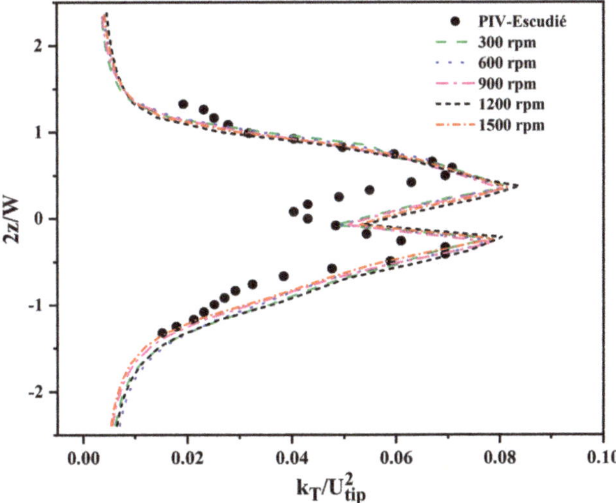

Figure 5. The turbulent kinetic energy normalized by U_{tip}^2.

3.3. Flow Rate Balance

The radial and axial flow rates through the control surfaces can be expressed as follows:

$$Q_r = 2\pi r \int_{z_1}^{z_2} U_r dz \qquad (8)$$

where z_1 = −20 mm, z_2 = 20 mm, r_1 = 0 or 55 mm, and r_2 = 55 or 85 mm. The radial pumping flow rate of the impeller (Q_r) is defined as the integral of the average radial velocity component over the entire impeller stream. Table 1 shows the volume flow rate balance around two control volumes that contain the impeller. One of the radial regions is r = 0~55 mm, the other is r = 55~80 mm, and both of them range from z = −20~20 mm. It could be seen that the differences between the flow in and flow out were within 1.5%, which proved the reliability of the simulation results [31].

Table 1. Flow rate balance around the impeller at 300 rpm, $-20 \leq z \leq 20$ (mm).

Region	Flow in (L/s)	Flow out (L/s)	Difference (%)
$0 \leq r \leq 55$ mm	0.3683	0.3735	1.4
$55 \leq r \leq 85$ mm	0.7474	0.7422	0.7

3.4. Trailing Vortices behind the Blades

Figure 6 shows the iso-surface of the Q criterion ($Q = 7 \times 10^4$ s^{-2}) behind each blade at different rotation times, which are coloured based on the velocity magnitude. The three-dimensional structure of the turbulent vortex was clear from the simulation results. The large-scale eddies embody themselves in the form of identifiable and organized distributions of the vorticity and are mainly responsible for the mixing performance [38,39]. The trailing vortices, one upper and one lower, formed at the rear of the blade and were attached to the end of the impeller disk. They unfolded along the blade outline, and then separated from the blade tip. There was a high-speed belt between the two vortices that lie in the bottom of the upper vortex and the top of the lower vortex. The fluid element is reduced by the vortex split and energy transfer to achieve small size micro mixing. The instantaneous structures of the trailing vortices accompanied by the vortex tube's winding, kink, fracture and reattachment were generated at different evolution times. The trailing vortices are associated with high shear rates and strong turbulent activity, and therefore they are essential to the mixing performance of the flow field [40]. The continuous movement and deformation of the vortex in the turbulent flow field could accelerate the energy dissipation and the loss of the mechanical energy, which could enhance the interaction between the flotation reagents and mineral particles to improve the separation efficiency of the coal fly ash flotation with the pulp preconditioning.

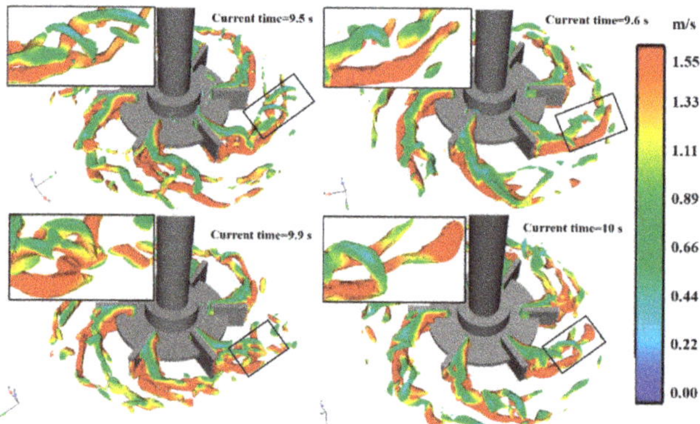

Figure 6. The instantaneous structure of the turbulent vortices.

3.5. Power Consumption under Different Rotation Speeds

The numerical simulation model used in this paper was able to predict the power consumption of the stirred tank. The power consumption (P) was calculated from the equation $P = 2\pi MN$, where M is the torque of the impeller, and N is the impellor's rotation speed [41]. The power number Np ($N_P = \frac{P}{\rho N^3 D^5}$, where ρ is the density of water and D is the impeller diameter) was used to check the validity of the LES model used in the simulation [42]. The power consumption and power number of the impeller under different rotation speeds are shown in Figure 7. The power consumption increased with the rotation speed, while the power number had a slight change. The power consumption had a large growth trend as it varied from 5.86 W to 736.06 W when the rotation speed changed from 300 rpm to 1500 rpm. Hudcova et al. also studied the influence of flow patterns on the power consumption and flow regime transition in a gas-liquid stirred vessel [43]. The average Np of the simulation results was found to be approximately constant at 4.65 for all the Reynolds numbers at the tested range, which was logical in comparison with that reported by Zadghaffari et al. of Np = 4.8 for six Rushton turbines [34]. Np represents the effective transmission of the energy to the processed liquid by the impellers [44].

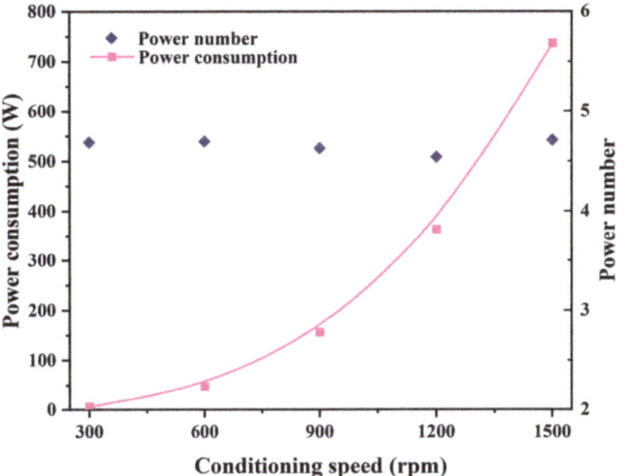

Figure 7. Power consumption and power number under different rotation speeds.

3.6. Turbulent Length Scale

The energy input agitated by the impeller rotation that dissipated into the stirred tank could accelerate the interaction between the mineral particles and flotation reagents. The smallest turbulent microscale of the energy dissipation of the energetic vortices is also called the Kolmogorov scale (η), which can be estimated by the dissipation rate of the turbulent kinetic energy, and can be expressed as

$$\eta = \left(\frac{v^3}{\epsilon} \right)^{1/4} \tag{9}$$

where v is the kinematic viscosity and ϵ is the dissipation rate of the turbulent kinetic energy. Most turbulent kinetic energy is consumed by the viscosity of the smallest eddy [45].

The Kolmogorov scale (η) at the position of $r/R = 1.07$ and $-2 < 2z/w < 2$ under different rotation speeds is plotted in Figure 8. Figure 8 shows that η decreases as the rotation speed increases. The Kolmogorov scale ranged from 11 μm to 87 μm in the jet flow when the rotation speed was 300 rpm, and then it narrowed to 3~46 μm at 1200 rpm and 3~38 μm at 1500 rpm. This magnitude is consistent with those previously reported [46,47]. The minimum Kolmogorov scale under the tested rotation speeds was approximately 3 μm, and it remained mostly unchanged when the rotation speed was beyond 1200 rpm. The energy that dissipated into the stirred tank could strengthen the interaction process between particles and reagents and improve the separation efficiency of the coal fly ash flotation.

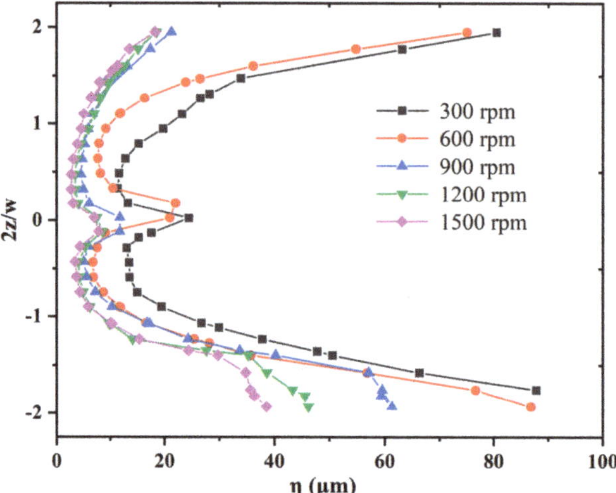

Figure 8. The Kolmogorov scale (η) under different rotation speeds.

4. Conditioning-Flotation Experiments

4.1. Materials and Methods

The coal fly ash samples were collected from a thermal power plant in Hunan, China. The particle size analysis according to the wet screening showed that −0.045 mm was the dominant size fraction with a yield of 50.72%, a loss on ignition of 12.14% and a carbon distribution of 48.40%. This showed that the flotation efficiency of fine particles was crucial in the separation of unburned carbon from coal fly ash. The contact angle of the fly ash sample was measured using an optical contact angle analysis system (DSA100, Krüss, Germany). It decreased quickly with the increase of the measurement time and was almost 0° at 2 s, as shown in Figure 9, which implied poor floatability of fly ash [48–50]. Light diesel oil and a 730-flotation reagent (prepared by mixing polyethylene glycol and 2-octanol at an appropriate ratio) were used as the collector and frother, respectively.

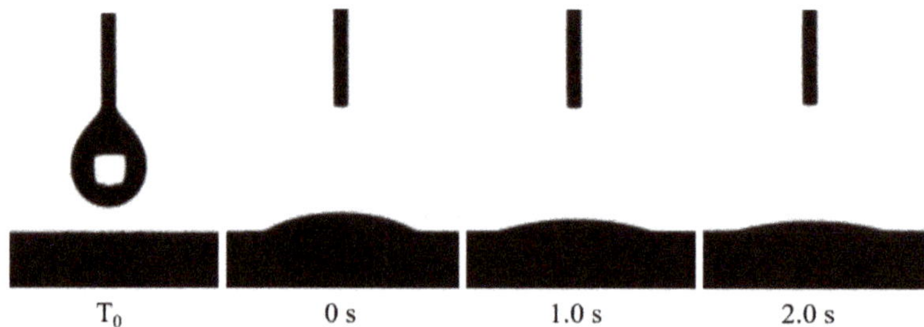

Figure 9. Contact angle of fly ash samples.

Stirred tanks are widely used for pulp preconditioning processing. Stirred tank was a standard cylindrical vessel, and the geometric dimensions were the same as those in the simulation (Figure 1). For the preconditioning tests, the height of the working fluid was equal to the diameter of the stirred tank. Then, the FA samples and collector were added into the stirred tank with a concentration of approximately 100 g/L and a dosage of 1000 g/t, respectively. The conditioning speeds were set as 300 rpm, 600 rpm, 900 rpm, 1200 rpm, and 1500 rpm at the conditioning time of 5 min (according to the preliminary tests). The pulp of 1 L was collected and prepared for the following flotation tests.

The flotation tests were carried out with a 1.0 L XFD lab-scale single flotation cell at an impeller speed of 1600 rpm. Then, the frother was added to the slurry and stirred for 1 min. Subsequently, the air with a flow rate of 1.67 L/min was introduced into the flotation cell. The total flotation time was 4 min. The collected concentrates and tailings were filtered, dried, and weighed for the following analysis.

The muffle furnace combustion method is used to measure the loss on ignition (LOI) of the concentrates and the tailings product [8,51]. The LOI is calculated as

$$\text{LOI}(\%) = \frac{W_b - W_a}{W_b} \times 100 \tag{10}$$

where W_b and W_a are the weight of the samples before burning and after burning, respectively.

The removal rate of unburned carbon (RUC) in the tailings is equal to the recovery rate of unburned carbon in the concentrates for the flotation tests and is computed a

$$\text{RUC}(\%) = \frac{LOI_C \times Y_C}{LOI_R} \times 100 \tag{11}$$

where LOI_C and Y_C are the LOI and yield of the concentrates, respectively, and LOI_R is the LOI of the raw fly ash.

4.2. Experimental Results and Discussion

4.2.1. Effect of the Conditioning Speed on the LOI and Yield of Flotation Tailings

The pulp preconditioning before flotation processing could promote the flotation behaviour of fine mineral particles. The effects of the condition speed on the LOI and the yield of the flotation tailings are shown in Figure 10. The LOI decreased as the conditioning speed increased. Then, it had a slight increase while the yield had a slight change. The yield of the flotation tailing ash varied from 66.87% to 78.69% at the tested range (300 rpm~1500 rpm). The LOI of the trailing ash decreased from 7.06% to 3.32% as the LOI of the feed coal fly ash samples was 12.72% when the conditioning speed increased from 300 rpm to 1200 rpm. The flotation tailing products, which had a lower LOI (less than 8%), could be used in the concrete according to the national standard GB/T 1596-2017. The probability of collision and attachment between the mineral particles and the flotation reagent increased with the turbulence fluctuation, which was caused by the increase of the conditioning speed, while the excessive fluctuation frequency would increase the possibility of detachment. Deglon concluded that increasing the level of agitation generally has a beneficial effect on the flotation rate of platinum ores [52].

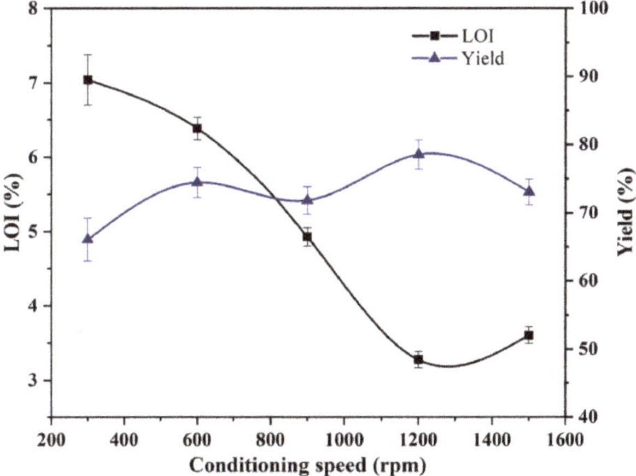

Figure 10. Effect of conditioning speed on the LOI and yield of concentrates.

4.2.2. Effect of the Conditioning Speed on the RUC of Flotation Concentrates

Figure 11 shows the effect of the conditioning speed on the recovery rate of unburned carbon (RUC) in the conditioning-flotation tests. It can be seen that the RUC under the tested conditioning speeds first increases and then decreases with the increase of the conditioning speed. The RUC reached the maximum of 80.89% at the conditioning speed of 1200 rpm. The UC collected from the FA could be used as some industrial materials or fed back to the boiler as fuel energy [53]. The optimal conditioning-flotation results were obtained with an LOI of 3.32%, a yield of 78.69% and an RUC of 80.89% when the conditioning speed was 1200 rpm. This was consistent with the LES results that the smallest turbulent scale (η) was obtained at the rotation speed of 1200 rpm, and the energy that dissipated into the stirred tank would not increase when the rotation speed exceeded 1200 rpm. Many authors claimed that energy transferred in the conditioning stage had a pronounced effect on the concentrate recovery, grade and flotation rate [16]. The preconditioning process with the energy input by the stirred tank could strengthen the interaction process between the particles and reagents, while the excessive energy input could not improve the flotation efficiency. The conditioning-flotation tests revealed the effect of the turbulence hydrodynamics due to the impeller rotation on the intensification of the coal fly ash flotation in the presented of preconditioning with a stirred tank.

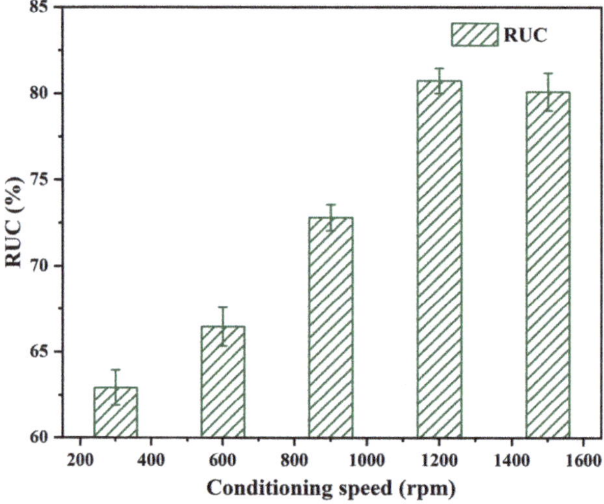

Figure 11. Effect of different conditioning speeds on the RUC of the flotation tests.

5. Conclusions

The pulp preconditioning using a stirred tank is important to strengthen the separation efficiency of coal fly ash in the flotation process. The turbulence characteristics and the turbulence length scale under different rotation speeds in the stirred tank were studied using large eddy simulation (LES). The conditioning-flotation tests were carried out to valid the effect of the turbulence flow field on the flotation strengthening by investigating the effect of the conditioning speed on the removal of unburned carbon from coal fly ash. The main conclusions are as follows.

The LES results were in good agreement with the previous experimental data. The turbulent kinetic energy, power consumption and instantaneous structure of trailing vortices at different evolution times were investigated in detail.

The turbulence length scale (η) decreased as the rotation speed increased and the minimum value of η was approximately 3 μm, and it remained mostly unchanged when the rotation speed exceeded 1200 rpm. The turbulent kinetic energy that dissipated into the stirred tank dominate the process of the intensification in stirred tank.

The conditioning-flotation tests showed that the optimal results were obtained with an LOI of 3.32%, a yield of 78.69% and an RUC of 80.89% at the conditioning speed of 1200 rpm. This was consistent with the simulation results, which showed the process strengthening effect of the turbulence hydrodynamics in the stirred tank on the flotation separation of unburned carbon from coal fly ash.

Author Contributions: Conceptualization, L.Y. and H.Z.; Methodology. X.Y.; Valid, L.Y., Z.Z. and X.Q.; Writing-Original Draft Preparation, L.Y.; Writing-Review & Editing, L.Y., X.Y. and H.Z.; and Visualization, H.Z.

Funding: This research was funded by the National Natural Science Foundation of China (Grant No. 51722405), and the Project Funded by the Priority Academic Program Development of Jiangsu Higher Education Institutions (PAPD).

Conflicts of Interest: The authors declare no conflict of interests.

References

1. Xu, Z.; Liu, J.; Choung, J.W.; Zhou, Z. Electrokinetic study of clay interactions with coal in flotation. *In. J. Miner. Process.* **2003**, *68*, 183–196. [CrossRef]
2. Hartuti, S.; Hanum, F.F.; Takeyama, A.; Kambara, S. Effect of Additives on Arsenic, Boron and Selenium Leaching from Coal Fly Ash. *Minerals* **2017**, *7*, 99. [CrossRef]
3. Li, S.; Wu, W.; Li, H.; Hou, X. The direct adsorption of low concentration gallium from fly ash. *Sep. Technol.* **2016**, *51*, 395–402. [CrossRef]
4. *ASTM D7348-13, Standard Test Methods for Loss on Ignition (LOI) of Solid Combustion Residues*; ASTM International: West Conshohocken, PA, USA, 2013.
5. *Fly Ash Used for Cement and Concrete*; PRC National Standard; Standardization Administration of the People's Republic of China: Beijing, China, 2017; GB/T 1596-2017.
6. Bartoňová, L. Unburned carbon from coal combustion ash: An overview. *Fuel Process. Technol.* **2015**, *134*, 136–158. [CrossRef]
7. Cameán, I.; Garcia, A.B. Graphite materials prepared by HTT of unburned carbon from coal combustion fly ashes: Performance as anodes in lithium-ion batteries. *J. Power Sources* **2011**, *196*, 4816–4820. [CrossRef]
8. Cao, Y.; Li, G.; Liu, J.; Zhang, H.; Zhai, X. Removal of unburned carbon from fly ash using a cyclonic-static microbubble flotation column. *J. S. Afr. I. Min. Metall.* **2012**, *112*, 891–896.
9. An, M.; Liao, Y.; Zhao, Y.; Li, X.; Lai, Q.; Liu, Z.; He, Y. Effect of frothers on removal of unburned carbon from coal-fired power plant fly ash by froth flotation. *Sep. Sci. Technol.* **2018**, *53*, 535–543. [CrossRef]
10. Peng, Y.; Grano, S. Effect of iron contamination from grinding media on the flotation of sulphide minerals of different particle size. *Int. J. Miner. Process.* **2010**, *97*, 1–6. [CrossRef]
11. Piñeres, J.; Barraza, J. Energy barrier of aggregates coal particle-bubble through the extended DLVO theory. *Int. J. Miner. Process.* **2011**, *93*, 14–20. [CrossRef]
12. Hao, J.; Gao, Y.; Khoso, S.; Ji, W.Y.; Hu, Y.H. Interpretation of Hydrophobization Behavior of Dodecylamine on Muscovite and Talc Surface through Dynamic Wettability and AFM Analysis. *Minerals* **2018**, *8*, 391.
13. Sun, W.; Xie, Z.J.; Hu, Y.H.; Deng, M.J.; Yi, L.; He, G.Y. Effect of high intensity conditioning on aggregate size of fine sphalerite. *T. Nonferr. Metals. Soc.* **2008**, *18*, 438–443. [CrossRef]
14. Feng, B.; Feng, Q.M.; Lu, Y.P. The effect of conditioning methods and chain length of xanthate on the flotation of a nickel ore. *Miner. Eng.* **2012**, *39*, 48–50. [CrossRef]
15. Yu, Y.X.; Cheng, G.; Ma, L.Q.; Huang, G.; Wu, L.; Xu, H. Effect of agitation on the interaction of coal and kaolinite in flotation. *Powder Technol.* **2017**, *313*, 122–128. [CrossRef]
16. Tabosa, E.; Rubio, J. Flotation of copper sulphides assisted by high intensity conditioning (HIC) and concentrate recirculation. *Miner. Eng.* **2010**, *23*, 1198–1206. [CrossRef]
17. Feng, D.; Aldrch, C. Effect of preconditioning on the flotation of coal. *Chem. Eng. Commun.* **2005**, *192*, 972–983. [CrossRef]
18. Pan, A.; Xie, M.H.; Li, C.; Xia, J.Y.; Chu, J.; Zhuang, Y.P. CFD Simulation of Average and Local Gas–Liquid Flow Properties in Stirred Tank Reactors with Multiple Rushton Impellers. *J. Chem. Eng. Jpn.* **2017**, *50*, 878–891. [CrossRef]
19. Vlček, P.; Kysela, B.; Jirout, T.; Fořt, I. Large eddy simulation of a pitched blade impeller mixed vessel-comparison with LDA measurements. *Chem. Eng. Res. Des.* **2016**, *108*, 42–48. [CrossRef]

20. Eggels, J.G. Direct and Large-eddy Simulation of Turbulent Fluid Flow Using the Lattice-Boltzmann Scheme. *Int. J. Heat Fluid* **1996**, *17*, 307–323. [CrossRef]

21. Bakker, A.; Oshinowo, L.M. Modelling of turbulence in stirred vessels using large eddy simulation. *Chem. Eng. Res. Des.* **2004**, *82*, 1169–1178. [CrossRef]

22. Zhou, G.; Kresta, S.M. Correlation of mean drop size and minimum drop size with the turbulence energy dissipation and the flow in an agitated tank. *Chem. Eng. Sci.* **1998**, *53*, 2063–2079. [CrossRef]

23. Smoluchowski, M.V. Versuch einer mathematischen theorie der koagulations kinetis kolloider losungen. *Z. Phys. Chem.* **1918**, *92*, 129–168.

24. Bakker, R.A. Micromixing in Chemical Reactors: Models, Experiments and Simulations. Ph.D. Thesis, TU Delft, Delft, The Netherlands, 1996.

25. Yoon, H.S.; Balachandar, S.; Man, Y.H. Large eddy simulation of passive scalar transport in a stirred tank for different diffusivities. *Int. J. Heat Mass Tran.* **2015**, *91*, 885–897. [CrossRef]

26. Zadghaffari, R.; Moghaddas, J.S.; Revstedt, J. Large-eddy Simulation of Turbulent Flow in a Stirred Tank Driven by a Rushton Turbine. *Comput. Fluids.* **2010**, *39*, 1183–1190. [CrossRef]

27. Yeoh, S.L.; Papadakis, G.; Lee, K.C.; Yianneskis, M. Large Eddy Simulation of Turbulent Flow in a Rushton Impeller Stirred Reactor with Sliding-deforming Mesh Methodology. *Chem. Eng. Technol.* **2004**, *27*, 257–263. [CrossRef]

28. Min, J.; Gao, Z.M. Large Eddy Simulations of Mixing Time in a Stirred Tank. *Chinese J. Chem. Eng.* **2006**, *14*, 1–7. [CrossRef]

29. Derksen, J. Assessment of Large Eddy Simulations for Agitated Flows. *Chem. Eng. Res. Des.* **2001**, *79*, 824–830. [CrossRef]

30. Wang, L.; Wang, Y.; Yan, X.; Wang, A.; Cao, Y.A. Numerical study on efficient recovery of fine-grained minerals with vortex generators in pipe flow unit of a cyclonic-static micro bubble flotation column. *Chem. Eng. Sci.* **2017**, *158*, 304–313. [CrossRef]

31. Wu, H.; Patterson, G.K. Laser-Doppler Measurements of Turbulent-flow Parameters in a Stirred Mixer. *Chem. Eng. Sci.* **1989**, *44*, 2207–2221. [CrossRef]

32. Escudié, R. Structure de L'hydrodynamique Générée par une Turbine de Rushton. Ph.D. Thesis, INSA de Toulouse, Toulouse, France, 2001. (In French)

33. Jaworski, Z.; Dyster, K.N.; Nienow, A.W. The effect of size, location and pumping direction of pitched blade turbine impellers on flow patterns: LDA measurements and CFD predictions. *Chem. Eng. Res. Des.* **2001**, *79*, 887–894. [CrossRef]

34. Zadghaffari, R.; Moghaddas, J.S.; Revstedt, J. A Mixing Study in a Double-Rushton Stirred Tank. *Comput. Chem. Eng.* **2009**, *33*, 1240–1246. [CrossRef]

35. Malik, S.; Lévêque, E.; Bouaifi, M.; Gamet, L.; Ghamri, N.; Simoëns, S.; El-Hajema, M. Shear improved smagorinsky model for large eddy simulation of flow in a stirred tank with a rushton disk turbine. *Chem. Eng. Res. Des.* **2016**, *108*, 69–80. [CrossRef]

36. Zhang, Y.H.; Yang, C.; Mao, Z. Large eddy simulation of liquid flow in a stirred tank with improved inner-outer iterative algorithm. *Chinese J. Chem. Eng.* **2006**, *14*, 47–55. [CrossRef]

37. Soos, M.; Kaufmann, R.; Winteler, R.; Kroupa, M.; Lüthi, B. Determination of maximum turbulent energy dissipation rate generated by a rushton impeller through large eddy simulation. *Aiche J.* **2013**, *59*, 3642–3658. [CrossRef]

38. Bakker, A.; Oshinowo, L.M.; Marshall, E.M. Chapter 31—The Use of Large Eddy Simulation to Study Stirred Vessel Hydrodynamics. In Proceedings of the 10th European Conference, Delft, The Netherlands, 2–5 July 2000; pp. 247–254.

39. Escudié, R.; Bouyer, D.; Liné, A. Characterization of trailing vortices generated by a rushton turbine. *Aiche J.* **2004**, *50*, 75–86. [CrossRef]

40. Derksen, J.; Akker, H. Large eddy simulations on the flow driven by a rushton turbine. *Aiche J.* **2010**, *45*, 209–221. [CrossRef]

41. Deglon, D.; Meyer, C. CFD modeling of stirred tanks: Numerical considerations. *Miner. Eng.* **2006**, *19*, 1059–1068. [CrossRef]

42. Bartels, C.; Breuer, M.; Wechsler, K.; Durst, F. Computational fluid dynamics applications on parallel-vector computers: computations of stirred vessel flows. *Comput. Fluids.* **2002**, *31*, 69–97. [CrossRef]

43. Hudcova, V.; Machon, V.; Nienow, A.W. Gas-liquid dispersion with dual Rushton turbine impellers. *Biotechnol. Bioeng.* **1989**, *34*, 617–628. [CrossRef] [PubMed]

44. Taghavi, M.; Zadghaffari, R.; Moghaddas, J.; Moghaddas, Y. Experimental and cfd investigation of power consumption in a dual rushton turbine stirred tank. *Chem. Eng. Res. Des.* **2011**, *89*, 280–290. [CrossRef]

45. Escudié, R.; Liné, A. A simplified procedure to identify trailing vortices generated by a rushton turbine. *Aiche J.* **2010**, *53*, 523–526. [CrossRef]

46. Escudié, R.; Liné, A. Experimental analysis of hydrodynamics in a radially agitated tank. *Aiche J.* **2003**, *49*, 585–603. [CrossRef]

47. Yeoh, S.L.; Papadakis, G.; Yianneskis, M. Numerical simulation of turbulent flow characteristics in a stirred vessel using the les and rans approaches with the sliding/deforming mesh methodology. *Chem. Eng. Res. Des.* **2004**, *82*, 834–848. [CrossRef]

48. Dey, S. Enhancement in hydrophobicity of low rank coal by surfactants-a critical overview. *Fuel Process. Technol.* **2012**, *94*, 151–158. [CrossRef]

49. Peng, Y.; Liang, L.; Tan, J.; Sha, J.; Xie, G. Effect of flotation reagent adsorption by different ultra-fine coal particles on coal flotation. *Int. J. Miner. Process.* **2015**, *142*, 17–21. [CrossRef]

50. Xia, W.; Yang, J.; Liang, C. A short review of improvement in flotation of low rank/oxidized coals by pretreatments. *Powder Technol.* **2013**, *237*, 1–8. [CrossRef]

51. Zhou, F.; Yan, C.; Wang, H.Q.; Zhou, S.; Liang, H. The result of surfactants on froth flotation of unburned carbon from coal fly ash. *Fuel* **2017**, *190*, 182–188. [CrossRef]

52. Deglon, D.A. The effect of agitation on the flotation of platinum ores. *Miner. Eng.* **2005**, *18*, 839–844. [CrossRef]

53. Hower, J.C.; Groppo, J.G.; Graham, U.M.; Wardb, C.R.; Kostovac, I.J.; Maroto-Valer, M.M.; Dai, S.F. Coal-derived unburned carbons in fly ash: A review. *Int. J. Coal Geol.* **2017**, *179*, 11–27. [CrossRef]

Case Report

Heavy Mineral Sands in Brazil: Deposits, Characteristics, and Extraction Potential of Selected Areas

Caroline C. Gonçalves and Paulo F. A. Braga *

Centre for Mineral Technology (CETEM), Av. Pedro Calmon 900, Cidade Universitária, Rio de Janeiro 21941-908, Brazil; carolinec.gon@gmail.com
* Correspondence: pbraga@cetem.gov.br; Tel.: +55-21-3865-7222

Received: 15 January 2019; Accepted: 6 March 2019; Published: 13 March 2019

Abstract: In Brazil, heavy mineral sand deposits are still barely exploited, despite some references to Brazilian reserves and ilmenite concentrate production. The goal of this project is to characterize and investigate the potential recovery of heavy minerals from selected Brazilian placer occurrences. Two areas of the coastal region were chosen, in Piaui state and in Bahia Provinces. In all samples, the heavy minerals of interest (ilmenite, monazite, rutile, and zircon) were identified by scanning electron microscopy (SEM) and X-ray diffraction (XRD) techniques and also quantified by X-ray fluorescence spectrometry (XRF) and Inductively Coupled Plasma Optical Emission Spectrometry (ICP-OES). The total heavy minerals (THM) in the Piaui samples were 6.45% and 10.14% THM, while the figure for the Bahia sample was 3.4% THM. The recovery test of the Bahia sample, using only physical separation equipment such as a shaking table and magnetic separator, showed valuable metallurgical recoveries at around or greater than 70% for each stage, and the final concentrate of pure ilmenite was composed of up to 60.0% titanium dioxide after the differential magnetic separation. Another aim is to compile accessible information about Brazilian heavy mineral main deposits complemented with a short economic overview.

Keywords: heavy mineral sands; Brazilian occurrences; SEM analysis; XRF analysis

---

1. Introduction

Although Brazil is considered the biggest producer of titanium dioxide in Latin America [1,2], mostly from ilmenite, studies about heavy minerals in the national territory are still scarce, limiting exploration works. The constant reference to heavy minerals such as ilmenite, zircon, and rutile as byproducts of rare earth element (REE) deposits, reveals the lack of information or a consolidated database about Brazilian heavy mineral sand deposits. The mention of defunct institutions as producers, and inaccurate grade-tonnage calculations of heavy mineral sands in international publications are common mistakes due to dispersed and hard-to-access information. Research aiming to identify new target areas and to compile accessible data seems to be a necessary step to plan future projects, considering the relevance of heavy mineral deposits in the international market.

In fact, heavy mineral sands are different than other commodities in terms of exploration, development, mining, and processing, but similar in the matter of importance to industry due to their relevant physical properties [3]. Heavy mineral sand deposits are generally voluminous and near to the surface, facilitating simple exploration techniques and open cast excavation [4]. Moreover, heavy mineral processing plants are currently well-established and highly mechanized, with considerable separation efficiency [4]. Usually, heavy minerals from beach sands are concentrated by physical methods, first involving gravity separation, followed by the combination of magnetic and electrostatic separation [5–8]. Flotation is also used [9,10], often as a subsequent and/or complementary stage, in an effort to achieve a cleaner product [11].

Globally, the leading producer of heavy minerals is Australia followed by South Africa [12]. The massive Australian participation in the international market has been consolidated since the 1960s [11]. Furthermore, according to Gosen et al. (2016) [4], China had the biggest production of ilmenite concentrate in 2014. In Brazil, however, the deposits of heavy mineral sands are largely not exploited, with the only active operation being the Guaju mine (Cristal Group) located in Paraiba Province. The main minerals processed from that mine are ilmenite, rutile, zircon, and kyanite. Likewise, the company Industrias Nucleares do Brasil (INB), in Rio de Janeiro Province, still produces ilmenite, zircon, rutile, and monazite concentrates for sale as is.

The importance of coastal heavy mineral deposits as sources of rutile, ilmenite, and zircon for industry, and their worldwide uses, have always been mentioned [13]. These deposits are a source of titanium and zirconium [4] as well as rare earth elements (REEs), and are often overlooked [14]. In Brazil, some found occurrences of monazite and its secondary products (ilmenite, zircon, and rutile) are along the coastline [15,16], and most studies have been specifically developed to find REE deposits (Figure 1). Practically all of them are located in the southeast region as described in Louwerse (2016) [17].

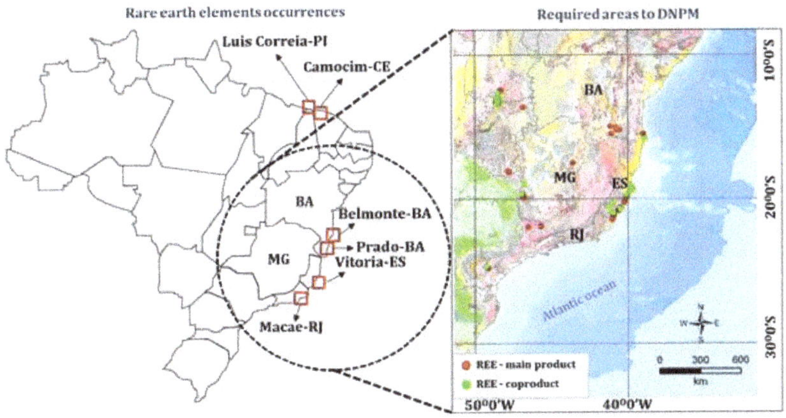

Figure 1. Rare earth elements (REEs)-bearing mineral occurrences along the Brazilian coastline and areas licensed by the National Department of Mineral Production (DNPM) for extraction of REEs. Adapted from [15,16].

As cited by CETEM (2014) [18], the presence of surface and submerged deposits of ilmenite, rutile, zircon, and monazite is along nearly the entire Brazilian littoral, from Para to Rio Grande do Sul. In these regions, including the northern portion, other profitable placer deposits of heavy mineral sands are/were being exploited, like in Sao Francisco de Itabapoana (Rio de Janeiro), Mataraca (Paraiba), and Prado (Bahia). However, according to Krishnamurthy and Gupta (2016) [19], mineral sand mining in Brazil is oriented to the thorium content of monazite, in contrast with other countries, where monazite is a secondary product rather than ilmenite, rutile, and zircon. In fact, in the past, exploration efforts for finding monazite deposits were connected with the intention of producing REE oxides for nuclear purposes [18]. INB for instance, which has mined and processed heavy minerals in the northern coastal region of Rio de Janeiro, is recognized for its monazite sand deposits. The Heavy Minerals Unit of Buena is the most relevant plant for monazite concentration, but has been decommissioned. Nowadays, monazite concentrates come from old mining activities and will likely be produced for two or three more years.

The production of titanium dioxide in Brazil started in 1971, and at present, heavy minerals such as ilmenite, rutile, kyanite, and zircon [20,21] are extracted at the Guaju mine [5] by the Cristal Group. The presence of monazite is also well known on the deposit [22,23]. Nevertheless, their reserves may be exhausted within a few years. Currently, some efforts for an independent national production of titanium mineral concentrates have been made by Rio Grande Mineraçao S. A., based on a forecast for growth of the Brazilian ilmenite consumption. They estimate an annual production of 323,000 tons of heavy mineral concentrates, 275,000 being ilmenite, 10,000 rutile, and 38,000 zircon [24].

A comparison of Brazilian heavy mineral resources and production with other countries is given every year by the National Department of Mineral Production (DNPM) [1]. Data compilation from 2009 to 2015 for ilmenite, rutile, and zircon are shown in Figure 2. Although Brazil has modest resources and production of zircon concentrate, in the national market of heavy mineral sands, ilmenite and rutile concentrates are considered the main products. Indeed, in more recent data [25], Brazilian ilmenite resources are still considered small, only about 5.6% of global resources. As described in Technical Report 36 [26], Brazilian production of titanium concentrates (ilmenite plus rutile) has oscillated considerably, with the most relevant decline since 1986 occurring in 2001, when production was only 64,450 tons. The highest titanium production, however, was in 2007 with 226,865 tons, and in 2015, the most recent data, was 81,000 tons [1].

Furthermore, the production of zircon, from 2003 to 2015, varied from 16,000 to 28,000 tons per year and the assigned resources decreased from 74 Mt in 2009 to approximately 22,6000 tons in 2015 (Figure 2). Placer deposits are considered the secondary source of zircon in Brazil, since primary sources are those related with intrusive alkaline rocks [1]. The monazite resources and production data are always included as one of the sources of REEs and their specific quantification is more complex. The preliminary calculated Brazilian monazite resources are 22 Mt [1]. The production of monazite concentrates, however, has significantly oscillated, falling from 1,173 tons in 2007 to 249 tons in 2010. More recent data showed a production of 600 tons in 2013 and no production in 2015 [1]. Although this was a small production, all monazite concentrate produced by INB in this period from 2003 to 2015 was exported [1].

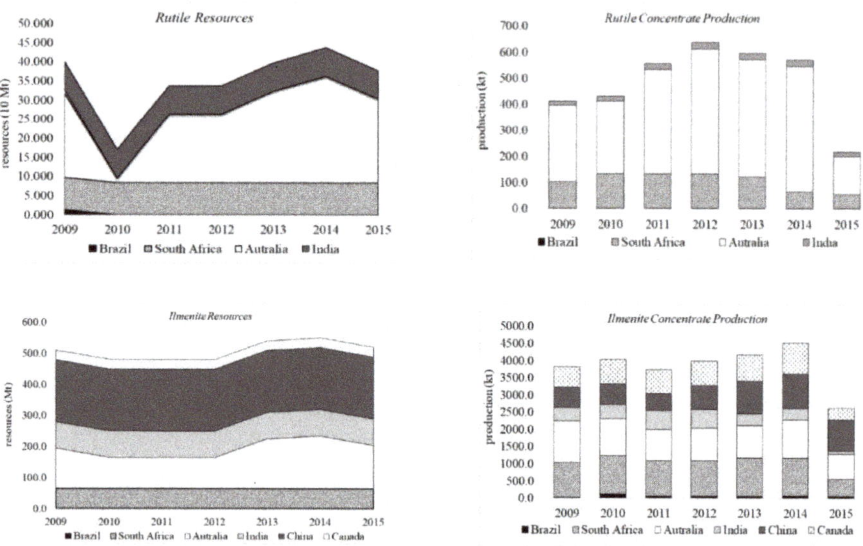

Figure 2. *Cont.*

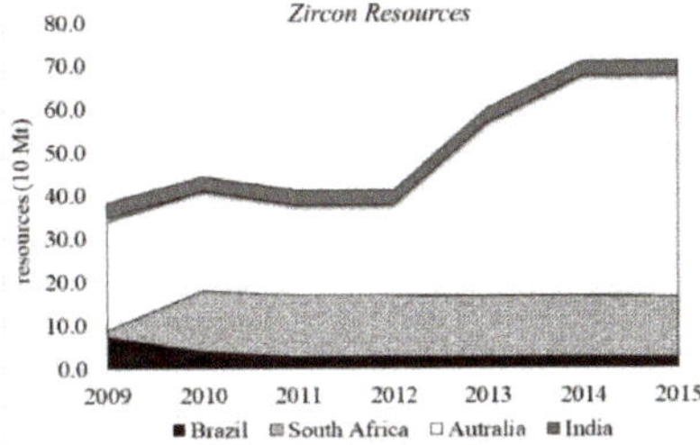
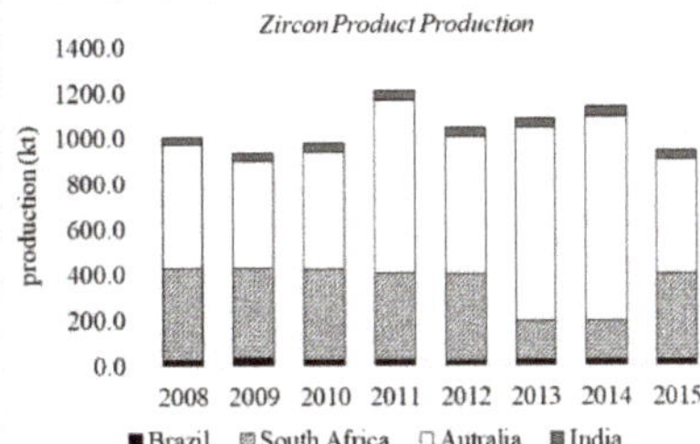

Figure 2. Brazilian resources and the production of ilmenite, rutile, and zircon concentrates, in comparison with the main global producers. Compiled from [1].

Therefore, a project for the characterization and recovery of heavy minerals from selected Brazilian placer occurrences was developed. The study was aimed at compiling accessible information about the main Brazilian deposits of heavy mineral sands [15,16,27–32], complemented by a short economic overview, highlighting the presence of ilmenite, zircon, monazite, and rutile. Two areas of the Brazilian coastal region were chosen for characterization studies.

2. Study of Potential Areas

2.1. Sampling Areas

Two potential areas were chosen based on previous studies [27,30] and recent exploration works [31]. These areas are located in Luis Correia/Piaui and Prado/Bahia. Three samples have been collected, two of them in Luis Correia (LC and LP) and one in Prado (PA) (Figure 3). In Luis Correia, the samples were collected by the grab sample technique, and due to a limited amount of collected material (at around 9.0 kg for each sample) and the lack of representativeness, LC and LP samples were characterized just in order to identify potential heavy minerals for future works. Conversely, around 0.5 tons of heavy mineral sands were collected in Prado (PA sample) by the trenching sampling method (at around 1.5 m depth). The spot was chosen supported by the local company's geochemical studies. The PA sample was also subjected to recovery tests.

The presence of heavy minerals in Luis Correia region was confirmed in the 1980s when the Samitre Company surveyed this occurrence and classified it as a medium deposit, with resources of approximately 1,500,000 tons [30]. According to the Brazilian Mineral Resources Company (2014) [29], Luis Correia district is classified as coastal deposits with fine sand. Moving dunes (Figure 3a,b) and paleodunes are composed mainly of selected quartz or quartz-feldspar sediments formed during the Quaternary Period [28].

Prado is also characterized by an accumulation of sediments of the Quaternary Period. Conversely, other authors [27] mentioned that the geological context of this area may be described by an accumulation of sediments from the Cretaceous to the Tertiary Periods associated with Archean rocks, which could have been a source of heavy minerals. Additionally, north Prado is in direct contact with Barreiras Group where sediments are classified as well as selected coarse sand (Figure 3c) [29]. About heavy mineral occurrences, Louwerse (2016) [17] cited that Prado region has monazite resources (approximately 4,464 tons) composed of 19.98% REE oxides, including the presence of xenotime and allanite minerals [17]. In fact, Cumuruxatiba and Alcobaça, near to Prado, have been sources of heavy minerals since the 18th century when monazite was sent to Europe [27].

Figure 3. Sampled areas: LP (**a**), LC (**b**), and PA (**c**).

2.2. Analytical Tools and Methods

Samples were characterized in three main parts: (1) the raw material; (2) sink dried aliquots separated by methylene iodide (d = 3.32 g·cm^{-3}) as a heavy medium and by a Frantz Isodynamic Separator (Model L-1, S.G. Frantz Co., Moorestown, NJ, USA) (transverse slope θ set at 15°), and (3) products and tailings recovered from test steps. Techniques and equipment are: optical microscopy, chemical analysis by X-ray fluorescence spectrometry (XRF, Axios Max Rh emission spectrometer, Malvern Panalytical Ltd., Royston, UK) and by inductively coupled plasma optical emission spectrometry (ICP-OES, Ultima 2 spectrometer, HORIBA Instruments Brasil, Ltda., Sao Paulo, Brazil); mineralogical analysis by X-ray diffraction (XRD, AXS D4 Endeavor co-emission diffractometer, Bruker, Karlsruhe, Germany) and by scanning electron microscopy (SEM, TM 3030 Plus tabletop microscope, Hitachi, Tokyo, Japan) with energy dispersive X-ray analysis (EDX); and density measurements by pycnometry (AccuPyc II 1340, Micromeritics, Norcross, GA, USA) with helium as analyzing gas. It is important to note that for XRF and XRD analysis, all samples tested were homogenized in longitudinal piles, and then quartered using a rotating laboratory sample divider. For XRD, representative aliquots of 3 g of each sample were ground with 10 mL of deionized water using a McCrone mill (Microne micronising mill, Glen Creston, Westmont, CA, USA) for 10 min, and after dried. For XRF, representative aliquots of 30 g were ground using a laboratory pulverizer (carbide ball mill, pulverisette, Fritsch, Idar-Olberstein, Germany) and then separated into 5 g aliquots using the same rotating laboratory sample divider.

2.3. Recovery Test

A pilot plant test using gravity and magnetic separation methods (Figure 4a) was performed on a 28.7 kg PA sample. Before all tests, a wet cleaning stage was carried out due to a great amount of organic material, which was partially responsible for the dark color (Figure 3c). The cleaned sample still had 3.0% of organic material measured in terms of loss on ignition (LOI) using a TGA-701 Thermogravimetric Analyzer (LECO Instrumentos Ltda., Rio de Janeiro, Brazil). Then, the cleaned sample was wet screened using a 297 μm aperture and the undersize (pulp with 35% solids) sent to a Super Duty Diagonal Deck shaking table (slope set at 10°). Furthermore, the heavy fraction was separated in an Inbras high intensity magnetic separator with a magnetic field strength of 4.5 T (Tesla). However, the nonmagnetic concentrate had to be separated again from light minerals by using a shaking table. In addition, the magnetic concentrate (about 800.0 g) was wet screened using a 212 μm aperture, and then each aliquot (oversize and undersize) was separated at the same Inbras high intensity magnetic separator, changing its belt speed (Figure 4b). The belt speed was adjusted to

50 rpm (rotations per minute) for each test, reaching 300 rpm maximum. Due to a higher magnetic susceptibility, minerals such as ilmenite should be recovered at the highest belt speed.

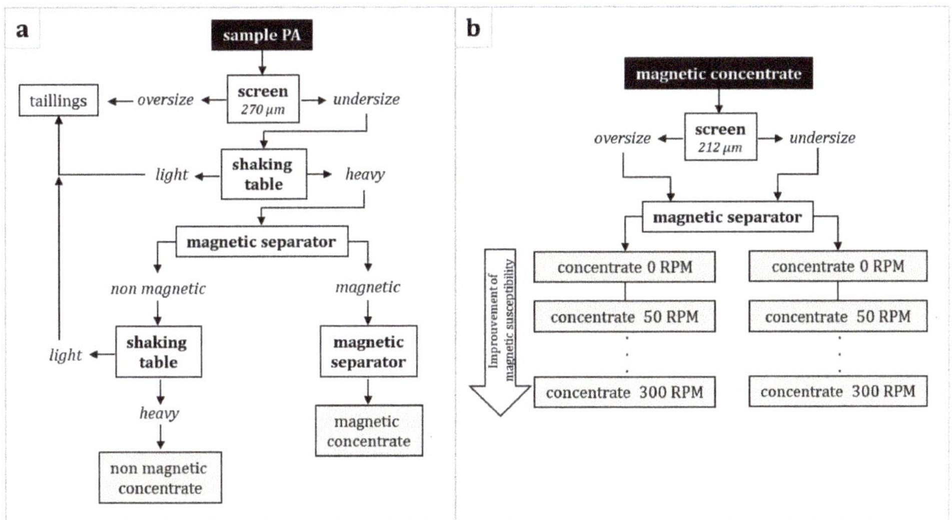

Figure 4. Main recovery test flow sheet (**a**) and differential magnetic separation flow sheet (**b**).

3. Results and Discussion

3.1. Characterization and Identification of Heavy Minerals

The samples' physical properties are given in Table 1. Specifically in this study, in order to facilitate characterization works, avoiding excessive contamination by silicates, total heavy mineral quantification was done using methylene iodide. Consequently, all minerals which have a density higher than 3.32 $g \cdot cm^{-3}$ were considered as heavy minerals (THM). Those heavier than 3.32 $g \cdot cm^{-3}$, which were recovered up to 1.3 T magnetic field strength by a Frantz Isodynamic Separator, were also considered as total magnetic heavy minerals.

Table 1. Classification of the three samples.

Property	LC	LP	PA
Total heavy minerals (THM) (%)	6.45	10.14	3.40
Total magnetic heavy minerals (%)	65.60	64.40	94.12
d_{80} passing size (µm)	190	400	730
Slimes (−45 µm) (%)	0.31	0.11	0.10

Based on the characterization analysis, both areas show the four minerals of interest, i.e., monazite (Mnz), rutile (Rt), zircon (Zrn), and ilmenite (Ilm), the latter mineral being the most abundant. Additionally, XRD and Rietveld quantification were performed. Using the software Bruker AXS Topas, (version 5), the phases were quantified. Their crystalline structures were identified using the following databases: Bruker AXS, Inorganic Crystal Structure Database (ICSD) [33], and Crystallographic Open Database (COD) [34]; refining variables were adjusted to those corresponding to the equipment standard. The result for the raw material demonstrated 0.7% ilmenite, 0.2% monazite, 0.3% rutile, and 0.2% zircon for LC; the LP sample was composed of 4.4% ilmenite, 1.6% monazite, 0.4% rutile, and 1.3% zircon. These two samples were also composed of minerals such as actinolite, epidote, muscovite, kyanite (Ky), staurolite (St), and leucoxene (Lcx). Details of some identified minerals from LP and LC are given in Figure 5. Due to the proximity of the sampled points, their mineralogical assemblages are quite similar. However, for the PA sample, although the Rietveld method allows the quantification of the amount of amorphous minerals [35], the results were not precise especially for minor minerals such as monazite and rutile due to the great presence of light minerals (higher than 96% in terms of mass distribution), and consequently, XRD interpretations were not considered.

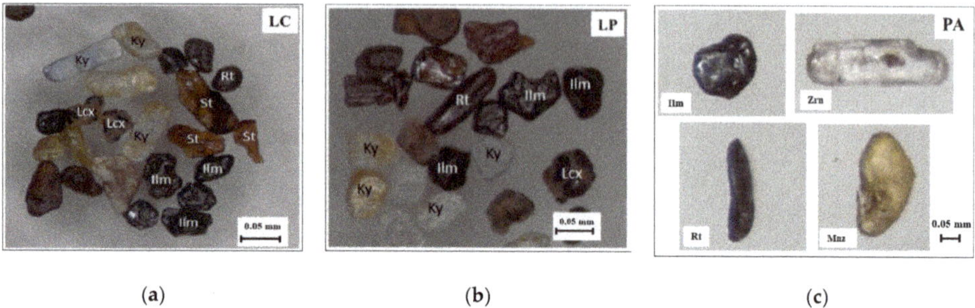

(a)   (b)   (c)

Figure 5. Examples of some heavy minerals identified via optical microscopy of LC (**a**), LP (**b**) and PA (**c**) samples. Note: Ilm: ilmenite, Ky: kyanite, Lcx: leucoxene, Mnz: monazite, Rt: rutile, St: staurolite and Zrn: zircon.

The separation by a heavy medium (methylene iodide) followed by a separation using a Frantz Isodynamic Separator allowed a better observation of those heavy minerals present in the samples. This stage provided important information about mineral alterations and inclusions, which can determine further beneficiation tests. An average of 65% by weight of LC and LP sink fractions were composed of magnetic heavy minerals. Ilmenite (Ilm) was the most abundant mineral in both samples and concentrated at 0.450 T fraction (Figure 6a); magnetite was also detected in this fraction. The ilmenite grains are extremely altered sometimes in pseudorutile and anatase (grain boundaries), as reported by [36], and also leucoxene (Figure 5a,b). These ilmenite grains showed selective alteration (Figure 6a) as well. The boundary alterations might be due to the reducing conditions of the geological environment [37]. However, although these alterations did not affect ilmenite recovery at the theoretical magnetic field strength at 0.450 T, ilmenite was also retained at 0.675 T. Ilmenite was usually found in pure form but a minor presence of manganese (Mn) has also been observed by EDX for both Luis Correia samples. Minerals such as monazite, stautolite, actinolite, and amphiboles were founded at 0.950 T fraction. In addition, the nonmagnetic fraction contains mostly prismatic and rounded zircon (Zrn) with kyanite (Ky) inclusions, and rounded and prismatic rutile (Rt) (Figure 6b). Despite the fact that zircon composition is not comparable to the theoretical due to the presence of niobium (Nb) and aluminum (Al), the hafnium (Hf) and zirconium (Zr) ratio are 0.04 for both samples as described elsewhere [38]. Al-silicates such as kyanite present in the nonmagnetic fractions of LC sample, for instance, are coarser than 74 μm, facilitating their separation by size from zircon; in other deposits this separation is more complicated [38] and a flotation stage is needed [10].

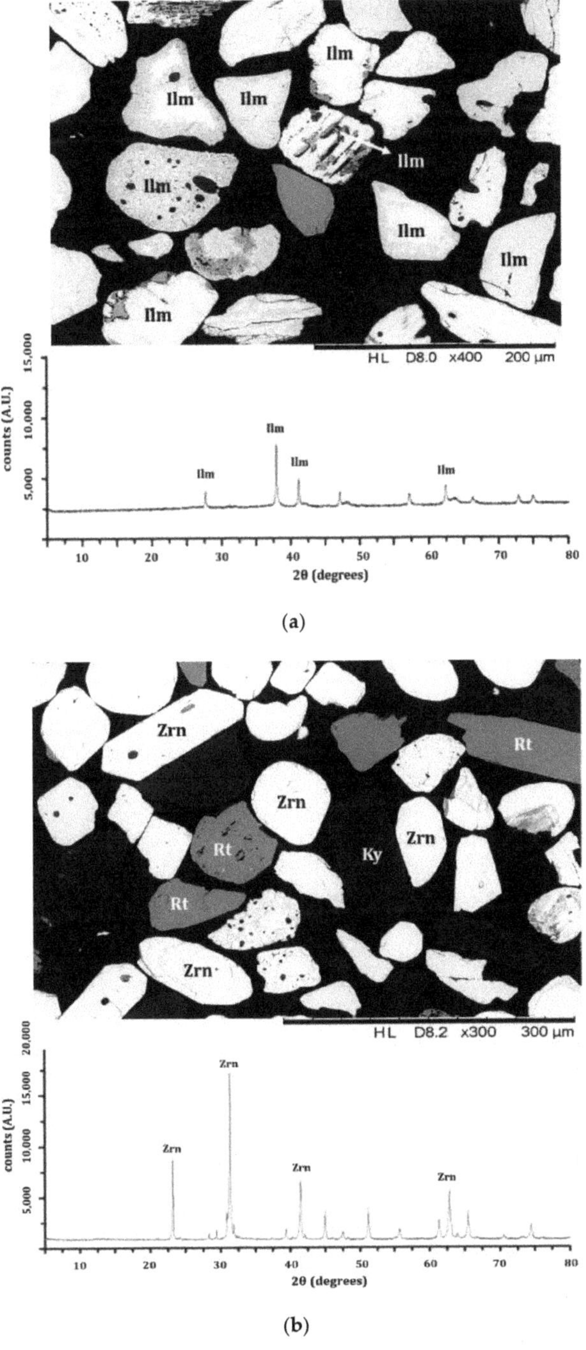

(a)

(b)

Figure 6. SEM images of LC heavy minerals retained at 0.450 T (**a**) and the nonmagnetic fraction (**b**) separated by a Frantz magnetic separator and their respective XRD spectra showing the main mineral of the fraction.

SEM image mapping of PA by element (Ti, P, Fe, Zr, and Si) of the sink fraction is shown below (Figure 7a). This image confirms the high presence of ilmenite rather than other heavy minerals such as zircon and monazite. In this analysis, no rutile was found. Monazite (Figure 7b) is the main mineral retained at 1.3 T and according to EDX data composed of REE Ce (24.4%), La (15.9%), Nd (6.1%), and Th (7.9%) on average. Monazite grain shapes are very variable and can vary from elongated to subrounded. In addition, in the nonmagnetic fraction, minerals such as zircon and rutile are mainly composed of 38.0% ZrO_2 and 53.1% SiO_2, and 89.3% TiO_2, respectively.

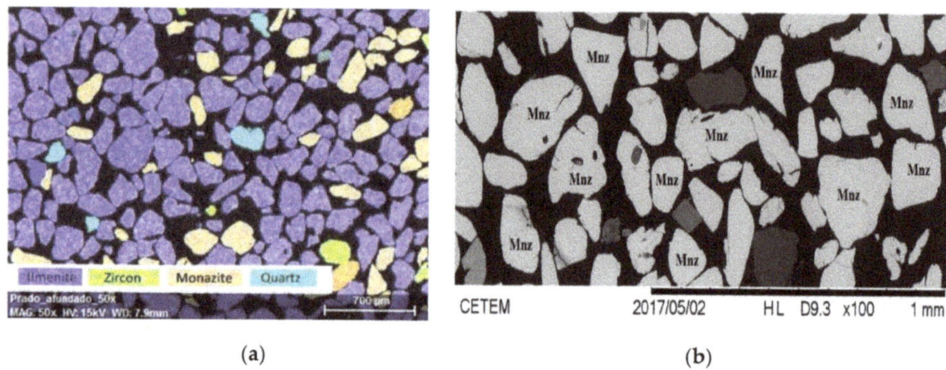

(a)   (b)

Figure 7. SEM image mapping of PA sink fraction (**a**) and SEM image of sink fraction retained at 1.3 T (**b**).

In PA heavy minerals retained at 0.675 T, it is possible to observe the presence of ilmenite, some zircon (inclusions), monazite, and xenotime. Although the latter is not abundant, it is also an interesting mineral since yttrium (Y) is an REE. According to EDX, its composition is 38.52% O, 37.34% Y, 22.96% P, and 1.17% Al. Ilmenite is the most abundant mineral in this fraction and is altered into anatase, especially along its boundaries and fractures (Figure 8a). Not just along its boundaries and fractures but EDX results showed that PA ilmenite might be considered an altered ilmenite since its Ti content is higher (34.85%) than the theoretical value (31.56%) [39]. Additionally, this Ti increase is observed followed by a decrease in the Fe percentage as suggested by Temple (1966) [40]. Another fact about PA ilmenite is the negligible amount of Mn in its composition. Nevertheless, this element was not quantified by EDX, the small Mn peak close to 6.0 kEV [41] can still be observed (Figure 8a). Some Fe remains in the composition of the ilmenite alteration, as can be seen in Figure 8b.

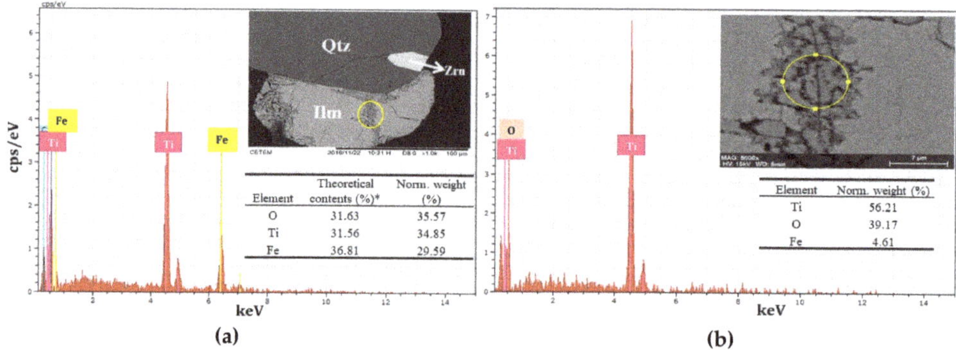

(a)   (b)

Figure 8. SEM images and EDX results for ilmenite (**a**) and details of its alteration (**b**). * Note: values from Dana (1974) [39].

Chemical analysis was also carried out. The head assay results of the oxides of interest are shown in Table 2. For PA, due to the low content of REEs (CeO_2, La_2O_3, and Nd_2O_3), thorium (Th), and zirconium (Zr), the chemical analysis by ICP-OES was carried out, showing 603.19 ppm of CeO_2, 570.91 ppm of La_2O_3, 450.42 ppm of Nd_2O_3, 107.51 ppm of ThO_2, and 456.77 ppm of ZrO_2.

Table 2. Selected major elements of whole rock sample according to X-ray fluorescence spectrometry (XRF) results.

Sample	Fe_2O_3 (%)	TiO_2 (%)	ZrO_2 (%)	REE + P_2O_5 (%)
LC	1.4	1.5	0.36	<0.1
LP	2.2	2.6	1.1	0.2
PA	0.7	1.5	<0.1	<0.1

Note: REE (CeO_2, La_2O_3, Nd_2O_3 + ThO_2).

Overall, it was possible to observe a significant difference between the Luis Correia samples and the Prado sample (Figure 9) in terms of in which particle size interval the oxides of interest (TiO_2, ZnO_2, Fe_2O_3, REE oxides, and P_2O_5) are. These oxides are preferentially concentrated at -150 $+53$ μm for LC (Figure 9a), at -106 $+53$ μm for LP (Figure 9b), and for PA at -297 $+106$ μm size fractions (Figure 9c). Specifically about PA REE oxides + P_2O_5, they tend to concentrate at -297 $+150$ μm, where almost 28.6% of the total mass contained 83.4% of the REE oxides and P_2O_5. Moreover, in LC, 31.39% P_2O_5 content is concentrated at -75 $+53$ μm and for LP this value is 46.57%, in terms of mass distribution.

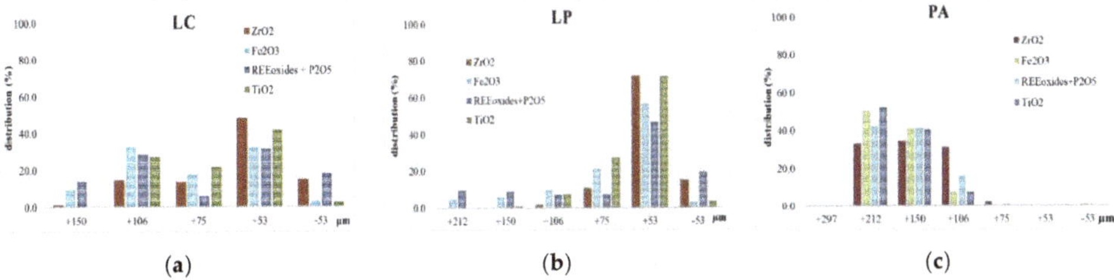

(a)   (b)   (c)

Figure 9. Selected major elements in terms of oxides (ZrO_2, Fe_2O_3, REE+P_2O_5 and TiO_2) of raw material by size fraction of LC (**a**), LP (**b**) and PA (**c**) samples, according to the XRF results.

3.2. Recovery Test of Bahia Sample

Mass and metallurgical balances of the recovery test are presented in Figure 10 as well as the particle distribution of the tested sample. The oxides considered for the process control were SiO_2 for quartz and other light minerals, TiO_2 for ilmenite and rutile, ZrO_2 for zircon, and P_2O_5 for monazite. The SiO_2 present in the zircon structure was discounted based on the average of the EDX values.

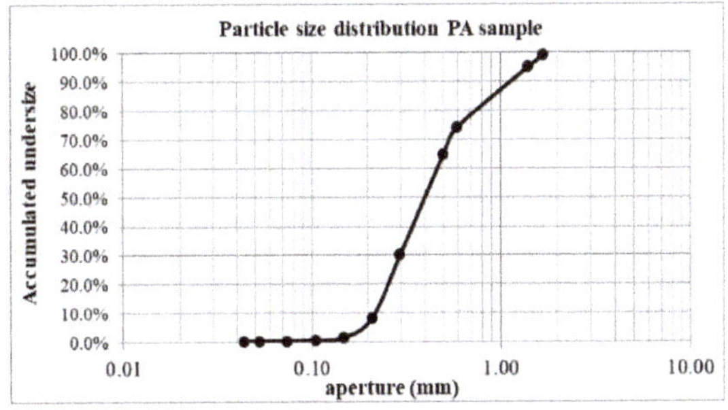

Figure 10. *Cont.*

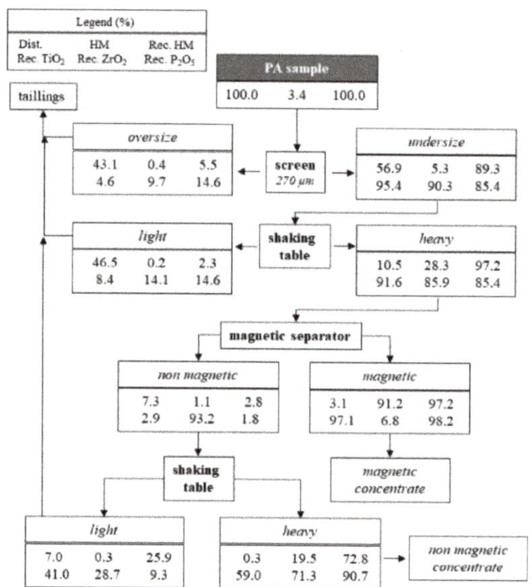

Figure 10. Particle size distribution of the PA sample (**a**) and mass and metallurgical balances (**b**).

The first gravity separation by shaking table shows values up to 85.0% for metallurgical recoveries of TiO_2, ZrO_2, and P_2O_5. The percentage of heavy minerals rises from 3.4% to 28.3%. After the magnetic separation stage, two bulk concentrates are obtained. Zircon is concentrated in the nonmagnetic concentrate, where 93.2% ZrO_2 is recovered. However, this fraction was still very contaminated with quartz (96.8% SiO_2 content due to quartz and other light minerals). Because of that, an additional shaking table cleaning stage was added in the process, obtaining a final nonmagnetic concentrate with 19.5% heavy minerals. This concentrate has 7.7% ZrO_2 with good ZrO_2 recovery (71.3%). Presumably, the titanium-bearing minerals such as rutile (Figure 5) are also in the nonmagnetic fraction due to its TiO_2 content (6.6%).

The magnetic concentrate is composed of 50.5% TiO_2 and 2.5% P_2O_5, and for the individualization of ilmenite and monazite, a differential magnetic separation was used, which related the distinct mineral magnetic susceptibilities with the belt speed of the magnetic separator (Figure 11). In this process, 0 rpm is a reference to nonmagnetic minerals still present in the magnetic concentrate (11.1% SiO_2 and 0.2% ZrO_2). The fractions tested were −270 μm, +212 μm, and −212 μm. The variations of fraction densities according to the belt speed are described in Figure 11a. These values were compared with theoretical densities of ilmenite (4.7 g·cm^{-3}), monazite (5.0 to 5.3 g·cm^{-3}) and quartz (2.65 g·cm^{-3}) [39]. Thus, it is possible to infer that monazite was concentrated at 150 rpm for the −212 μm fraction and at 200 rpm for the +212 μm fraction, whereas ilmenite was recovered at 100, 150, and up to 200 rpm for the +212 μm fraction, and at 150 rpm and up to 200 rpm for the −212 μm fraction. Consequently, the finer material (−212 μm fraction) has more minerals with low magnetic susceptibility, namely monazite, which is in accordance with Figure 9. The ilmenite final concentrate is considered that recovered at 250 and 300 rpm, and according to the chemical analysis by XRF (Figure 11b), TiO_2 content is slightly higher than 60.0%. This value is consistent with those found for the composition of PA ilmenite, suggesting the presence of altered ilmenite with a lower Fe content and higher Ti content, as verified in the characterization studies (Figure 8a). The P_2O_5 content reaches values of almost 20.0% at 150 rpm for the −212 μm fraction, while for the same belt speed the content was 10.0% for the coarser fraction (+212 μm) (Figure 11b). Overall, in both size fractions, SiO_2 content decline significantly (less than 5.0% in each material recovered from 50 to 300 rpm) and the average ZrO_2 content is 0.66% at 0 rpm. This result demonstrated the possibility of selective separation between ilmenite and monazite by varying the belt speed of the magnetic separator.

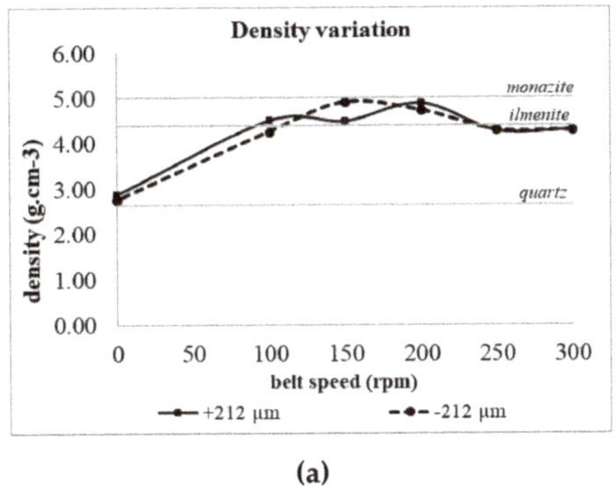

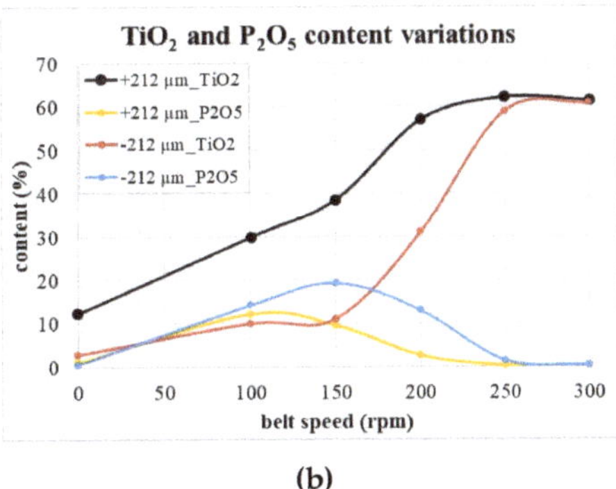

(a)   (b)

Figure 11. Density variations of the material recovered in different belt speeds (**a**), and variation of selected oxide (TiO$_2$ and P$_2$O$_5$) contents for +212 µm and −212 µm size fractions according to the magnetic separator's belt speeds (**b**).

4. Conclusions

This study shows a compilation of data of Brazilian heavy mineral sands. The Brazilian deposits of heavy minerals have been known since the 1940s, with the beginning of monazite concentrate production by INB. Some of these deposits are located along the southeast Brazilian coastline, especially in Rio de Janeiro and Espirito Santo Provinces. However, they are not widely explored, being restricted nowadays to the Brazilian northern region at Guaju mine (Cristal Group) in Paraiba Province, where the reserves may be exhausted within a few years. Only one new project, to extract ilmenite in Sao José do Norte, Rio Grande do Sul Province, was found.

The other question is that the constant reference to heavy minerals such as ilmenite, zircon, and rutile as byproducts of REE deposits gives the impression of a lack of information or a consolidated database in their respect. Normally these sources are restricted to private companies or dispersed among public entities. Despite the existence of references to Brazilian resources in many reports around the world, investments in this sector are low.

In the two studied areas (Piaui and Bahia), the heavy minerals of interest (ilmenite, monazite, rutile, and zircon) were identified by SEM and XRD analyses. The total heavy minerals of the two Piaui samples, LC and LP, were 6.45% and 10.14%, respectively, and for the Bahia sample, the figure was 3.4%. The head assays of the main oxides, by XRF, showed contents of 1.5%, 2.6%, and 1.5% of TiO$_2$ for the LC, LP, and PA samples. LP had the highest content of ZrO$_2$ (1.1%) and REEs (CeO$_2$, La$_2$O$_3$, Nd$_2$O$_3$, and ThO$_2$), and P$_2$O$_5$ (0.2%). For PA, due to the low contents of REEs, thorium (Th), and zirconium (Zr), chemical analysis by ICP-OES was carried out, showing 603.19 ppm of CeO$_2$, 570.91 ppm of La$_2$O$_3$, 450.42 ppm of Nd$_2$O$_3$, 107.51 ppm of ThO$_2$, and 456.77 ppm of ZrO$_2$. However, the THM of these occurrences is still quite low when compared with other deposits, and just reinforces the small Brazilian participation in this market (Figure 2). Investments in exploration works for heavy mineral sands could change this scenario.

The recovery test of the Bahia sample, using physical separation equipment such as a shaking table and a magnetic separator, showed valuable metallurgical recoveries at around or greater than 70% for each stage (Figure 10), demonstrating a reliable test execution. The final concentrate of ilmenite is composed of up to 60.0% titanium dioxide after differential magnetic separation. These concentrates obtained for +212 µm at 250 and 350 rpm, and −212 µm at 300 rpm could even be commercialized since their contents satisfy specifications of the Brazilian market (53% TiO$_2$ and <0.1% P$_2$O$_5$) [42]. These physical methods proved to be efficient to selectively separate ilmenite from monazite.

Author Contributions: C.C.G. performed all characterization analysis and mineral processing tests. P.F.A.B. contributed with the design of the mineral processing. Both C.C.G. and P.F.A.B. authors contributed to the final version of the manuscript. P.F.A.B. supervised the project.

Funding: This research received no external funding.

Acknowledgments: The authors acknowledge the Center for Mineral Technology (CETEM), which permitted this project development, and also the National Council for Scientific and Technological Development (CNPq) for the research grant; the Mineral Processing Laboratory of the University of São Paulo (LTM) for some tests, and the G-4 Esmeralda company for the PA sample.

Conflicts of Interest: The authors declare no conflict of interest.

References

1. National Department of Mineral Production (DNPM). Sumário Mineral. c2010, c2011, c2012, c2013, c2014, c2015, c2016. Available online: http://www.anm.gov.br/dnpm/publicacoes/serie-estatisticas-e-economia-mineral/sumario-mineral (accessed on 15 November 2018).
2. Filho, P.C.S.; Serra, O.A. Terras raras no Brasil: Histórico, produção e perspectivas. *Química Nova* **2014**, *37*, 12.
3. Jones, G. *Mineral Sands: An Overview of the Industry*; Iluka Company: Perth, Australia, 2008; p. 26.
4. Gosen, B.S.V.; Bleiwas, D.I.; Bedinger, G.M.; Ellefsen, K.J.; Shah, A.K. Coastal deposits of heavy mineral sands: Global significance and US resources. *Min. Eng.* **2016**, *68*, 36–43.
5. Sampaio, J.A.; Luz, A.B.; Alcantera, R.M.; Araújo, L.S.L. *Minerais Pesados Millennium. Usinas de Beneficiamento de Minérios no Brasil*; Sampaio, J.A., Luz, A.B., Lins, F.F., Eds.; Center for Mineral Technology: Rio de Janeiro, Brazil, 2001; p. 233.
6. Schnellrath, J.; Monte, M.B.M.; Veras, A.; Júnior, H.R.; Figueiredo, C.M.V. *Minerais Pesados INB. Usinas de Beneficiamento de Minérios no Brasil*; Sampaio, J.A., Luz, A.B., Lins, F.F., Eds.; Center for Mineral Technology: Rio de Janeiro, Brazil, 2001; p. 189.
7. Rosental, S. *Terras Raras, Rochas e Minerais Industriais Usos e Especificações*; Luz, A.B., Lins, F.F., Eds.; Center for Mineral Technology: Rio de Janeiro, Brazil, 2005; p. 727.
8. Laxmi, T.; Srikant, S.S.; Rao, D.S.; Rao, R.B. Beneficiation studies on recovery and in-depth characterization of ilmenite from red sediments of badlands topography of Ganjam District, Odisha, India. *Int. J. Min. Sci. Technol.* **2013**, *23*, 725–731. [CrossRef]
9. Routray, S.; Rao, R.B. Beneficiation and Characterization of Detrital Zircons from Beach Sand and Red Sediments in India. *J. Miner. Mater. Charact. Eng.* **2011**, *10*, 1409–1428. [CrossRef]
10. Routray, S.; Laxmi, T.; Rao, R.B. Alternate Approaches to Recover Zinc Mineral Sand from Beach Alluvial Placer Deposits and Bandlands Topography for Industrial Applications. *Int. J. Mater. Mech. Eng.* **2013**, *2*, 80–90.
11. Tranvik, E.; Becker, M.; Palsson, B.I.; Franzidis, J.P.; Bradshaw, D. Towards cleaner production–Using floatation to recover monazite from a heavy mineral sands zircon waste stream. *Miner. Eng.* **2017**, *101*, 30–39. [CrossRef]
12. Moore, P. Heavy Minerals. *Int. Min.* **2011**, *1*, 7.
13. Worobiec, A.; Stefaniak, E.A.; Potgieter-Vermaak, S.; Sawlowicz, Z.; Spolnik, Z.; Grieken, R.V. Characterization of concentrates of heavy minerals sands by micro-Raman spectrometry and CC-SEM/EDX with HCA. *Appl. Geochem.* **2007**, *22*, 2078–2085. [CrossRef]
14. Mudd, G.M.; Jowitt, S.M. Rare earth elements from heavy mineral sands: Assessing the potential of a forgotten resource. *Appl. Earth Sci.* **2016**, *125*, 107–113. [CrossRef]
15. Silveira, F.V.; Chemale, L.T. *Minerais Estratégicos: Avaliação do Potencial de Terras Raras do Brasil*; Serviço Geológico do Brasil–CPRM: Brasília, Brazil, 2012; p. 38.
16. Chemale, L.T. *Avaliação do Potencial de Terras Raras do Brasil*; Serviço Geológico do Brasil–CPRM: Brasília, Brazil, 2013; p. 23.
17. Louwerse, D. Rare Earth Element Deposits and Occurrences within Brazil and India–Indicating and Describing the Main REE Deposits and Occurrences and Their Potentialities. Bachelor's Thesis, Delft University of Technology, Delft, The Netherlands, 2016; p. 62.

18. Center for Mineral Technology (CETEM). *Exploração de Terras Raras em São Francisco do Itabapoana (RJ) Afeta Meio Ambiente, Banco de Dados de Recursos Minerais e Territórios: Impactos Humanos, Socioambientais e Econômicos*; Center for Mineral Technology: Rio de Janeiro, Brazil, 2013.

19. Krishnamurthy, N.; Gupta, C.K. *Extractive Metallurgy of Rare Earths Second Edition*; CRC Press: Boca Raton, FL, USA, 2016; p. 839.

20. Sabedot, S.; Sampaio, C.H. Caracterização de zircões da Mina Guaju (PB). *Revista Escola de Minas* **2002**, *55*, 5. [CrossRef]

21. Sabedot, S. Caracterização Tecnológica e Beneficiamento Mineral Para a Produção de Concentrado de Zircão de Alta Qualidade, Partindo de Pré-Concentrado e Concentrado de Zircão de Baixa Qualidade, Produzidos na Mina do Guaju (PB). Ph.D. Thesis, Universidade Federal do Rio Grande do Sul, Porto Alegre, Brazil, 2004; p. 114. (In Portuguese)

22. Ferreira, K.R.S. Caracterização do Concentrado de Ilmenita Produzido na Mina do Guaju, Paraíba, Visando Identificar Inclusões de Monazita e Outros Contaminantes. Master's Thesis, Universidade Federal do Rio Grande do Sul, Porto Alegre, Brazil, 2006; p. 64. (In Portuguese)

23. Ferreira, K.R.S.; Sabedot, S.; Sampaio, C.H. Evaluation of monazite presence in ilmenite concentrates produced at Guaju Mine (PB). *Revista Escola de Minas* **2007**, *60*, 669–673. [CrossRef]

24. Rio Grande Mineração. *Relatório de Impacto Ambiental Projeto Retiro*; CPEA—Consultoria, Planejamento e Estudos Ambientais Ltda and HAR Engenharia e Meio Ambiente Ltda: São José do Norte, Brazil, 2014; p. 37.

25. U.S. Geological Survey. *Mineral Commodity Summaries 2017*; U.S. Geological Survey: Reston, VA, USA; U.S. Department of Interior: Washington, DC, USA, 2017; p. 206.

26. Santos, J.F. *Technical Report 36: Titanium Profile*; J. Mendo Consultoria: Brasília, Brazil, 2010; p. 34.

27. Noguti, I.; Feitosa, J.A. *Potencial de Minerais Pesados na Região de Valença-Itacaré, Bahia—Amostragem, Caracterização e Tratamento do Minério*; Compania Bahiana de Pesquisa Mineral: Salvador, Brazil, 1970; p. 18.

28. Andrade, A.C.S.; Dominguez, J.M.L. Informaçoes geológico-geomorfológicas como subsídios à análise ambiental: Exemplo da planície costeira de Caravelas-Bahia. *Boletim Paranaense de Geociências* **2002**, *1*, 8. [CrossRef]

29. Brazilian Mineral Resources Company (CPRM). *Mapa geológico do estado do Piauí*; Serviço Geológico do Brasil: Brasília, Brazil, 2014; p. 1.

30. Moraes, A.M. *Diagnóstico e Diretrizes Para o Setor Mineral do Estado do Piauí—Parte I Caracterização Geral do Estado*; Ministério de Minas e Energia and Governo do Estado do Piauí: Teresina, Brazil, 2004; p. 170.

31. Mendes, J.C. Monazita e outros minerais pesados da região de Prado-Caravelas, no sul da Bahia. In Proceedings of the III Seminário Brasileiro de Terras Raras, Rio de Janeiro, Brazil, 27 November 2015; p. 14.

32. Cavalcanti, V.M.M. *Plataforma Continental: A Última Fronteira da Mineração Brasileira*; National Department of Mineral Production (DNPM): Brasília, Brazil, 2011; p. 96.

33. ICSD—Inorganic Crystal Structure Database. Available online: https://http://www2.fiz-karlsruhe.de/icsd_home.html (accessed on 14 December 2017).

34. Crystallography Open Database. Available online: https://http://www.crystallography.net/cod/ (accessed on 14 December 2017).

35. Neumann, R.; Schneider, C.L.A.; Neto, A.A. *Caracterização Tecnológica de Minérios: Parte II. Tratamento de Minérios*; Luz, A.B., Sampaio, J.A., França, S.C.A., Eds.; Center for Mineral Technology: Rio de Janeiro, Brazil, 2010; p. 85.

36. Nair, A.G.; Babu, D.S.S.; Damodaran, K.T.; Prabhu, C.N. Weathering of ilmenite from Chavara deposit and its comparison with Manavalakurichi placer ilmenite, southwestern India. *J. Asian Earth Sci.* **2009**, *34*, 115–122. [CrossRef]

37. Weibel, R.; Friis, H. Alteration of opaque heavy minerals as a reflection of the geochemical conditions in depositional and diagenetic environments. *Dev. Sedimentol.* **2007**, *58*, 277–303.

38. Reyneke, L.; van der Westhuizen, W.G. Characterization of a heavy mineral-bearing sample from India and the relevance of intrinsic mineralogical features to mineral beneficiation. *Miner. Eng.* **2001**, *14*, 1589–1600. [CrossRef]

39. Dana, J.D. *Manual de Mineralogia*; Livros Técnicos e Científicos: Rio de Janeiro, Brazil, 1974; Volume 2, p. 287.

40. Temple, A.K. Alteration of ilmenite. *Econ. Geol.* **1966**, *61*, 695–714. [CrossRef]

41. Braccini, S.; Pellegrinelli, O.; Kramer, K. Mössbauer, X-Ray and magnetic studies of black sand from the Italian Mediterranean Sea. *World J. Nucl. Sci. Technol.* **2013**, *3*, 91–95. [CrossRef]
42. Cristal. Minérios. Available online: https://www.cristal-al.com.br/nossos-produtos/minerios/ (accessed on 28 January 2019).

Article

Mineralogy and Pretreatment of a Refractory Gold Deposit in Zambia

Xu Wang [1], Wenqing Qin [1,*], Fen Jiao [1,*], Congren Yang [1], Yanfang Cui [1], Wei Li [1], Zhengquan Zhang [1] and Hao Song [2]

[1]   School of Mineral Processing and Bioengineering, Central South University, Changsha 410083, China
[2]   College of Engineering, Drexel University, Philadelphia, PA 19102, USA
*   Correspondence: qinwenqing369@126.com (W.Q.); jfen0601@126.com (F.J.)

Received: 4 June 2019; Accepted: 27 June 2019; Published: 1 July 2019

Abstract: The technological mineralogy of a gold deposit located in North-Western province of Zambia was carried out by using X-ray fluorescence spectroscopy (XRF), X-ray diffraction spectroscopy (XRD), and scanning electron microscope (SEM). The results showed that gold was highly dispersed in gold-bearing minerals such as pyrite, arsenopyrite, and some gangues in the form of natural gold and electrum. The gold grade in the mineral was 15.96 g/t and the particle size distribution of gold was extremely uneven. Most of the gold particles were less than 10 μm and wrapped with gold-bearing minerals, making it difficult to achieve liberation during grinding. According to the characteristics of the refractory gold deposit, the gravity–flotation combined beneficiation process was used to recover the liberated coarse gold and the fine gold in the sulphides. The closed-circuit experiments obtained excellent indicators. The grade and recovery of gold in the gravity separation concentrates reached 91.24 g/t and 57.58%, respectively. The grade and recovery of gold in the flotation concentrates were 49.44 g/t and 33.36%, respectively. The total recovery of gold was 90.94%. The gravity–flotation combined beneficiation pretreatment process provided a feasible method for the refractory gold ore and ensured the effective recovery of gold.

Keywords: gold deposit; mineralogy; gravity–flotation combined; pretreatment

---

1. Introduction

It is well known that gold is a precious metal and has been used as money, store of value, and jewelry for many centuries. The abundance of gold in the earth's crust is only three parts per billion [1]. With the rapid depletion of easy-to-treat gold deposits and in order to ensure the sustainability of resources, more attention has been paid to research on the recovery and utilization technology of refractory gold deposits [2,3].

The gold mineral should be pretreated before smelting due to the low grade of gold in the ore. The pretreatment process of gold mainly includes gravity separation, flotation, microbial oxidation, etc. [4,5]. Gold occurs usually in its native form or hosts in sulphides (especially pyrite, arsenopyrite, and chalcopyrite), silicate, carbonate, and oxide minerals [1,6,7]. Silver is generally associated with gold and it is usually recycled with gold [8,9].

The best pretreatment process is ultimately determined by the geological formation process and mineralogy of the gold deposit [10–13]. In general, liberated native gold is recovered by low cost gravity separation, especially for coarse gold [14,15]. Common gravity separation equipment includes shaking table, spiral chute, and jigging concentrator. The effectiveness of gravity separation equipment is closely related to the size and shape characterizations of gold particles [16–18]. The finer the grain size of gold particles, the more difficult they are to recover [18,19]. In sulphidic refractory gold ores, gold is highly disseminated and locked up in sulphide minerals, such as pyrite and arsenopyrite, in the form of micro or sub-micro particles, and is often recovered with these sulphide minerals by flotation [8,20]. Thiol collectors such as xanthates and dithiophosphates are the most widely used flotation reagents in the gold mining industry, which increase the difference of hydrophobicity between sulfide and gangues [21,22]. Copper sulfate and sulfuric acid can activate pyrite and arsenopyrite and enhance their floatability so as to improve recovery of the gold deposit [23]. Gold-bearing pyrite and arsenopyrite account for a considerable proportion of refractory gold-bearing sulfide ores. Gold can be recovered through bulk flotation or selective flotation processes according to the distribution of gold in pyrite and arsenopyrite [24,25]. The selective flotation of arsenopyrite from pyrite can be realized by adding a magnesium–ammonium mixture depressant [26]. As we all know, arsenic is a highly toxic substance that causes great harm to the human body and the environment. If arsenic pyrite enters the gold smelting process, the arsenic is enriched and the flow direction is difficult to control, making it more difficult to handle. If the associated relationship between gold and arsenopyrite is not tight and does not affect the gold recovery rate, the best method for removing arsenic is to discharge it into the tailings through a selective flotation process to ensure that arsenic is relatively stable and at a lower state [27–29]. Cyanide leaching is a traditional and methodically developed process, which is effective for extracting gold from refractory ores [30]. However, cyanide has gradually been replaced by other lixiviants such as glycine and thiourea because of its strong toxicity and serious environmental damage [31–33]. Nowadays, more and more new biological approaches are being studied to recover gold from refractory gold-bearing sulfide ore [5,34–36]. Although biometallurgy is environmentally friendly, it has not been applied on a large scale because its production cycle is too long and the environmental requirements of bacteria are strict [37,38].

A gold deposit containing pyrite and arsenopyrite invested in by Chinese companies is located in North-Western province in Zambia. The gold is pre-enriched by the method of beneficiation, and the obtained gold concentrate is transported back to China to recover gold by metallurgical method. In order to reduce subsequent transportation costs and metallurgical costs, it is desirable to pre-enrich the gold ore using a suitable pretreatment process. The gold grade in the raw ore is relatively high, as shown in Table 1. Systematic laboratory study of the mine was not conducted due to the local conditions. In the actual production process, regardless of the pretreatment process using gravity separation of shaking table or flotation, the gold recovery cannot achieve the expected indicators. At the same time, the high arsenic content in gold concentrates increases the environmental pressure of subsequent treatment processes. Therefore, this paper aims to characterize the high-sulfur and high-arsenic gold deposit and understand the chemical element composition, the mineralogical composition, and intercalation of minerals, especially the state of gold occurrence by process mineralogy. According to the results of process mineralogy, the reasons for the low efficiency of beneficiation and the possibility of removing arsenic from gold concentrate are analyzed, and a suitable pretreatment process is proposed to recover the refractory gold ore reasonably and efficiently.

Table 1. Chemical analysis of the raw ore (mass fraction, %).

S	Fe	Cu	Zn	As	Sb	Pb	Ag *	Au *
7.91	15.44	0.05	0.33	6.50	0.12	0.42	123.64	15.96

* The units of Au and Ag are g/t.

2. Materials and Methods

2.1. Mineral Samples

The samples used in this study were obtained from an arsenic-containing sulfur-bearing gold deposit in North-Western province, Zambia. The representative sample taken was mixed proportionally with the ore samples at each mining point. A bulk sample of 150 kg crushed to −2 mm was homogenized and split into 500 g representative sub-samples. The results of chemical analysis of raw ore are shown in Table 1. The content of gold in the ore was 15.96 g/t.

2.2. Methods

2.2.1. Mineralogical Study

The contents of different elements of the ore were tested by X-ray fluorescence (XRF, S0902724, Rigaku, Tokyo, Japan). The mineralogical composition and content of the ore were identified by X-ray diffraction (XRD, S0902240, Rigaku, Tokyo, Japan). The mineral paragenesis and size distribution were analyzed by scanning electron microscope (SEM, JSM−6360LV, JEOL, Tokyo, Japan), energy dispersive spectrometer (EDS, EDX-GENESIS, Ametek, Berwyn, PA, USA), electron microprobe analysis (EMPA, 1720H, Shimadzu, Japan), polarizing microscope (LEICA DM4500P, Germany), and other conventional analysis.

2.2.2. Gravity Separation and Flotation

Samples of 500 g with a particle size of 100% passing 2 mm were ground in a closed stainless steel Ø240 × 90 mm^2 XMQ conical ball mill (Wuhan Exploring Machinery Plant, Wuhan, China) at a slurry concentration of 66% (mass concentration). Gravity separation tests were carried out on a LY-1100 × 500 shaking table (Wuhan Exploring Machinery Plant, Wuhan, China). Flotation experiments were carried out using an XFD series single flotation cell (Jilin Exploring Machinery Plant, city, China) with volumes of 1.5 L, 1 L, and 0.5 L for rougher, scavenger, and cleaner flotation, respectively. Flotation experiments were conducted at a solid content of 25–40%. The flotation reagents were added to the cell and stirred for several minutes before rougher, scavenger, and cleaner flotation.

2.3. Reagents

Flotation experiments were carried out using sodium silicate, copper sulfate, and carboxymethocel (CMC) as regulators and terpilenol as frother. The collectors contained sodium butyl xanthate (SBX) and ammonium dibutyl dithiophosphate (ADD). All flotation reagents were acquired from Hunan Mingzhu flotation reagents limited company (Zhuzhou, China). All of these reagents were industrial grade and were diluted in water to a concentration of 1–5% before being added to the flotation tank.

3. Results and Discussion

3.1. Composition and Content of Ore

The modal mineralogy of the ore was mainly green schist. The key minerals were chlorite, mica, talc, and so on. Major alterations were silicification and pyrite mineralization. The results of chemical composition of the ore determined by XRF are listed in Table 2. It can be seen that the other valuable metals in the ore were Ag, Zn, and Pb. The main non-metallic element was Si. Figure 1 shows the XRD pattern of raw ore. The XRD quantitative analysis results indicated that the contents of quartz, talc, chlorite, pyrite, feldspar, mica, and arsenopyrite in the ore were 40.59%, 14.97%, 11.75%, 6.53%, 13.56%, 5.32%, and 13.56%, respectively. Gangue minerals were mainly quartz, chlorite, feldspar, talc, and mica.

Table 2. Element analysis of samples by XRF (mass fraction, %).

Mg	Al	Si	S	K	Ca	Ti
3.921	4.785	31.145	4.076	0.824	0.321	0.133
V	**Cr**	**Mn**	**Fe**	**Ni**	**Cu**	**Zn**
0.018	0.232	0.02	8.783	0.069	0.04	0.294
As	**Sr**	**Ag**	**Sb**	**Pb**		
4.99	0.005	0.011	0.124	0.388		

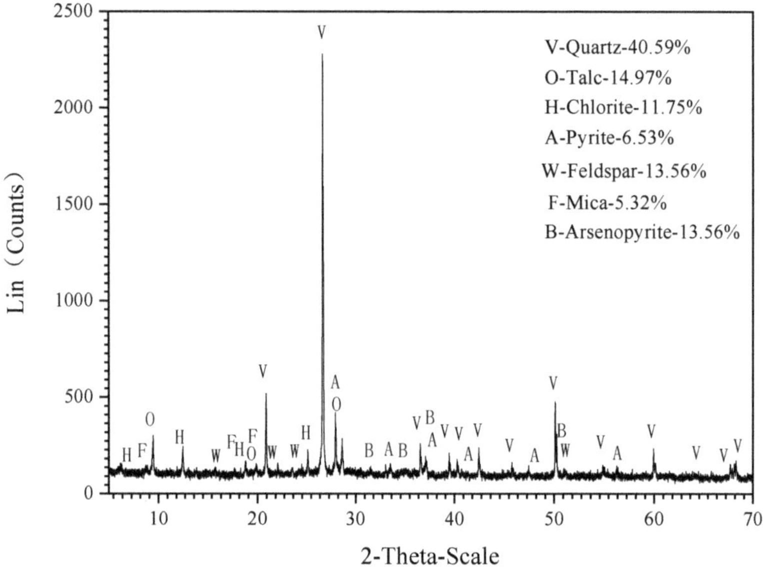

Figure 1. XRD pattern of raw ore.

The samples with a particle size of 73% passing 74 μm were used for phase analysis. Table 3 shows the result of chemical phase analysis of gold in the ore. It demonstrates that gold mainly existed in the form of native gold or was distributed in pyrite, arsenopyrite, and other oxides. Gold in native gold and sulphides accounted for 88.82%.

Table 3. Chemical phase analysis of gold in ore.

Phase	Native Gold	Gold in Pyrite and Arsenopyrite	Gold in other Sulphides	Gold in Oxides	Other	Total
Content (g·t^{-1})	10.82	2.90	0.42	1.67	0.11	15.92
Distribution (%)	67.96	18.22	2.64	10.49	0.69	100.00

3.2. Occurrence of Main Metallic Minerals

3.2.1. Gold

Gold mainly appeared in the form of natural gold and electrum in the ore. As the amount of silver in the gold increased, the color of gold under the microscope changed from golden yellow to bright yellow. Gold mainly occurred in the form of enclosing gold, fracture gold, and intergranular gold in pyrite, arsenopyrite, sphalerite, jamesonite, and gangues.

Gold in Pyrite

Pyrite was closely related with gold and was the main gold-bearing mineral. Gold was more likely to appear in the better crystallized pyrite. It could be seen that most of the gold was enclosed in pyrite by irregular, granular, strip, and a small amount of gold was distributed in cracks and holes in pyrite under the microscope. Most gold particles had a particle size of less than 10 μm. Very few large particles of gold reached 50 to 60 μm, and fine particles below 1 μm of gold particles accounted for a certain proportion. It was difficult to dissociate gold particles by grinding. Figure 2 shows images of gold in pyrite under reflected light.

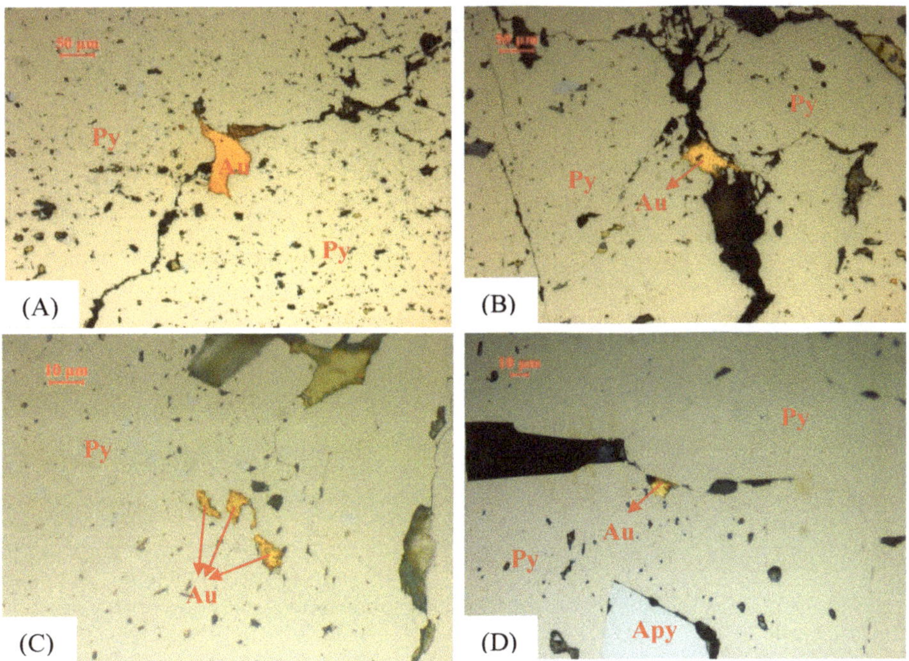

Figure 2. Images of gold in pyrite in reflected light: (**A**) enclosing gold in pyrite; (**B**) intergranular gold in pyrite; (**C**) enclosing gold in pyrite; (**D**) fracture gold in pyrite (Py—pyrite; Apy—arsenopyrite; Au—gold).

Figure 2A,B shows that gold with larger particles embedded in pyrite in an irregular shape and some fine gold particles with sizes between 3 μm and 10 μm were distributed around the large gold particles under the microscope. The long diameter of the larger gold particles in Figure 2A was 60 μm. Figure 2C,D shows the enclosing gold and fracture gold of the finer particles in the pyrite, respectively.

Figure 3A,B shows the SEM image and EDS analysis of large particle gold in Figure 2A,B, respectively. The results of EDS indicated that the content of Au in the gold particles was 98.02% in Figure 2A and the Au content was 72.20% and the Ag content in the gold particles was 27.80% in Figure 2B.

Gold in Arsenopyrite

Arsenopyrite was also the main gold-bearing mineral and it was closely related with gold. The distribution of gold in arsenopyrite was similar to that in pyrite. Most of the fine-grained gold with a particle size of less than 10 μm was highly dispersed in arsenopyrite, as shown in Figure 4A. The long diameter of the inclusive gold, as seen in Figure 4A, was about 10 μm. Figure 4B shows the large grain of gold distributed in the arsenopyrite hole. When the ore was crushed and ground, the hole may have been split and the gold was dissociated. Figure 5 shows SEM images and EDS analysis of intergranular gold in arsenopyrite. The result indicated that the Au content of gold particles was 72.27%, and the Ag content was 27.73%. The As content was 38.79%, S content 24.29%, and Fe content 36.92% of arsenopyrite around gold particles.

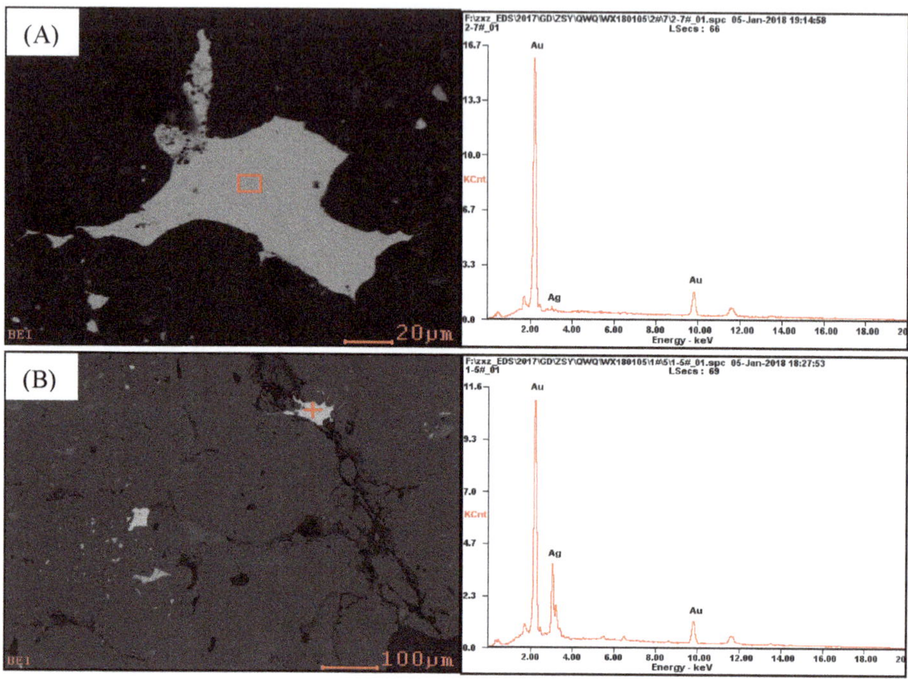

Figure 3. SEM images and EDS analysis of gold in pyrite: (**A**) enclosing gold in pyrite; (**B**) intergranular gold in pyrite.

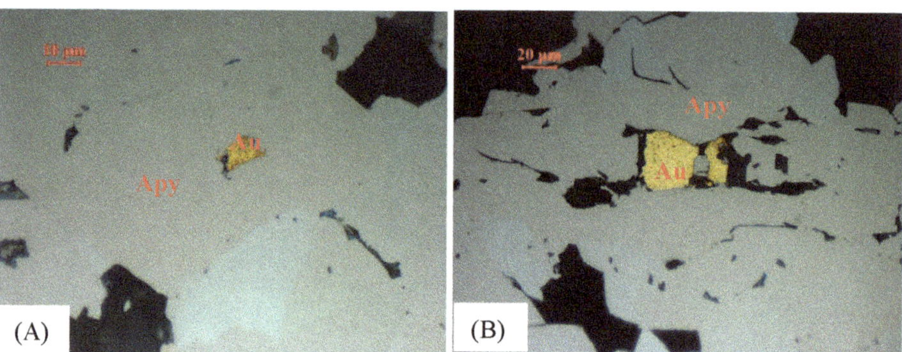

Figure 4. Images of gold in pyrite in reflected light: (**A**) enclosing gold in arsenopyrite; (**B**) hole gold in arsenopyrite (Apy—arsenopyrite; Au—gold).

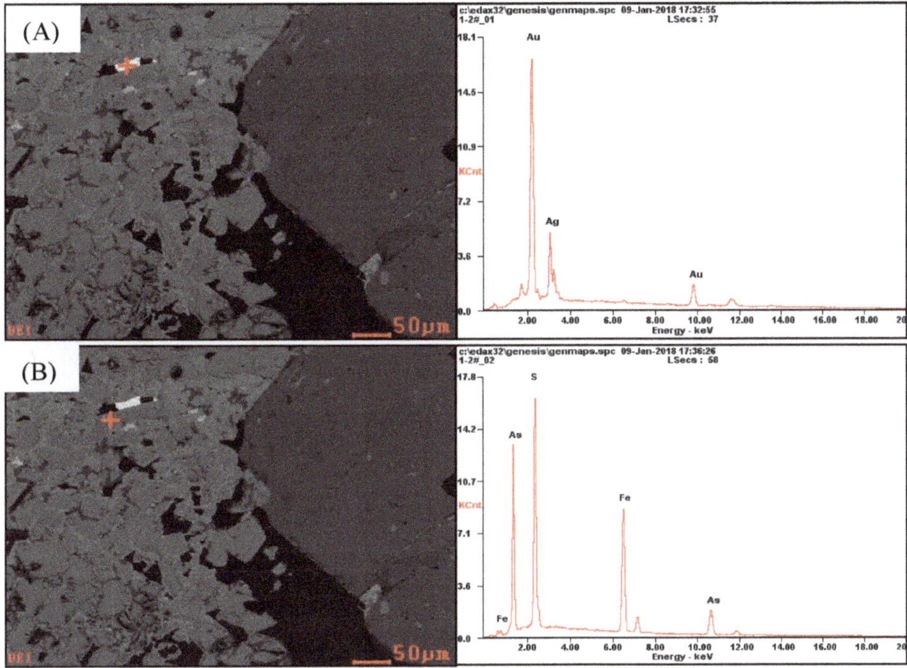

Figure 5. SEM images and EDS analysis of intergranular gold in arsenopyrite: (**A**) enclosing gold in arsenopyrite; (**B**) arsenopyrite.

Gold in other Minerals

In addition to the gold distributed in pyrite and arsenopyrite, a small amount of gold was distributed in other sulphides and gangues. Figure 6 shows images of gold in other minerals in reflected light. Gold was mainly distributed at the edge of galena, jamesonite, and sphalerite. The gold in the gangue was distributed in a granular or irregular shape. The coarse gold particles in gangue were about 40 μm, and the fine particles were 3–5 μm. Gold particles in the holes of gangue might have fallen off during grinding.

3.2.2. Pyrite and Arsenopyrite

Pyrite and arsenopyrite were the main sulfides in the ore. The pyrite particle size distribution in the ore was uneven, and it existed in the form of self-crystal, semi-self-crystal, and other crystals. There were many holes and cracks inside the pyrite particles. Figure 7A,B indicates that the ore contained coarse-grained arsenopyrite and fine-grained arsenopyrite. The coarse-grained arsenopyrites were cemented together to form a massive aggregate, and fine-grained arsenopyrites were cemented together to form a dense aggregate. The pyrite and arsenopyrites were closely related to each other in the ore. As shown in Figure 7D, the fine-grained arsenopyrite cemented pyrite formed a dense structure, making it difficult to separate by grinding. The intergranular pores of arsenopyrite were often cemented with sphalerite and galena. At the same time, arsenopyrite and pyrite were also distributed in other sulfide minerals.

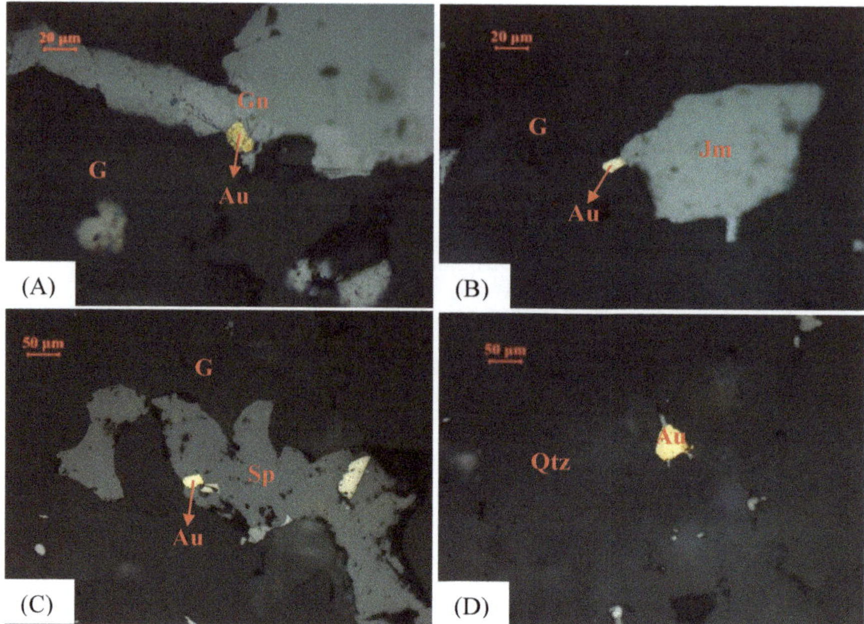

Figure 6. Images of gold in other minerals in reflected light: (**A**) gold associated with galena; (**B**) gold associated with jamesonite; (**C**) gold associated with sphalerite; (**D**) gold associated with quartz (Au—gold; Gn—galena; Jm—jamesonite; Sp—sphalerite; Qtz—quartz; G—gangue).

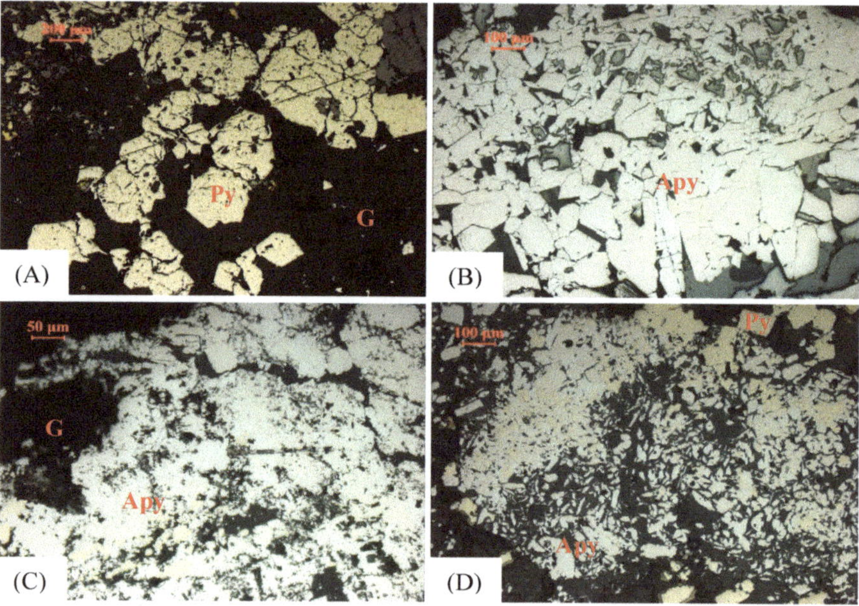

Figure 7. Images of pyrite and arsenopyrite in reflected light: (**A**) pyrite; (**B**) coarse arsenopyrite aggregate; (**C**) fine arsenopyrite aggregate; (**D**) an aggregate of pyrite and arsenopyrite (Py—pyrite; Apy—arsenopyrite; G—gangue).

3.2.3. Other Sulphides

The metallic elements Zn, Pb, Sb, and Cu in ore mainly occurred in the form of sphalerite, galena, jamesonite, and chalcopyrite. Figure 8A shows an image of an aggregate of sphalerite, jamesonite, and chalcopyrite, and Figure 8B shows an image of galena in reflected light of the microscope. The sphalerite was closely associated with chalcopyrite, and the galena ore was closely associated with jamesonite in the ore. Galena and jamesonite were usually in the form of symbionts with uneven grain size distribution in the ore. Some arsenopyrite was distributed in jamesonite in granular form.

Figure 8. Images of other sulphides in reflected light: (**A**) aggregate of sphalerite, jamesonite, and chalcopyrite; (**B**) image of galena (Sp—sphalerite; Ccp—chalcopyrite; Jm—jamesonite; Gn—galena).

3.3. Strategies to Recover Gold

The results of the mineralogy study showed that gold in the ore mainly existed in the form of natural gold and electrum. Pyrite and arsenopyrite were the main gold-bearing minerals, followed by sphalerite, galena, jamesonite, and some gangue minerals. Gold particles of 5–10 µm were most distributed in the ore, followed by the particle size less than 1 µm, which accounted for a certain distribution rate. The largest particle size of about 100 µm accounted for a very small part. The results of EDS indicated that the Au content of gold particles varied from 30 to 98%. Gangue minerals were mainly quartz, chlorite, feldspar, talc, and mica. Most of the fine-grained gold distributed in pyrite and arsenopyrite was mostly completely encapsulated, and it was difficult to dissociate. In addition, there was some gold in granular, irregular form distribution in quartz and other gangue minerals. The associated relationship between pyrite and arsenopyrite was also extremely close.

According to the results of mineralogy study, gold particles had a wide range of particle size distribution. The natural gold in gangue and the coarse-grained gold dissociated from sulphide ore could be recovered by gravity separation. However, the fine-grained gold associated with sulphides was difficult to be enriched and recovered by gravity separation. This might be the reason for the low efficiency of gravity separation in actual production. The gold associated with sulphides or fine-grained gold could be enriched into sulphide concentrate to the maximum extent by flotation. However, flotation has a poor recovery effect on coarse-grained gold. At the same time, the layered silicate minerals such as chlorite and talc had good natural floatability, and they were easily floated with the sulfide ore, which might lead to a decline in concentrate grade [39]. This might be the reason for the poor efficiency of flotation in actual production. Since some fine-grained grade gold was highly dispersed in arsenopyrite, and the associated relationship between pyrite and arsenite was extremely close, the removal of arsenic during the beneficiation process would reduce the recovery rate of gold. The nature of the ore determined that the gold mine did not have the innate conditions for arsenic removal during the beneficiation process. Therefore, the gold mine was a refractory gold deposit containing arsenic and sulfur.

In this refractory gold deposit, gold particles that were dissociated from the gangue and sulfide monomers after grinding could be recovered by gravity beneficiation. The fine-grained gold still encased in pyrite and arsenite was recovered by bulk flotation. Therefore, the gravity–flotation combined beneficiation pretreatment process might be a reasonable way to recover gold from ore. Figure 9 shows the flow chart of the gravity–flotation combined beneficiation.

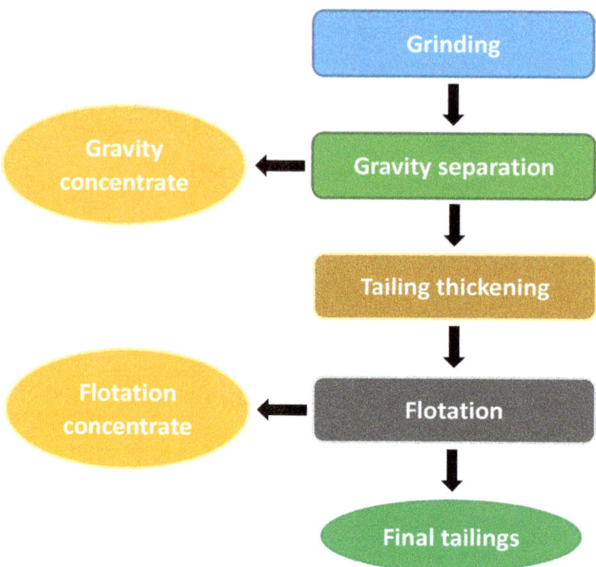

Figure 9. The flow chart of gravity–flotation combined beneficiation.

3.4. Gravity Separation

According to the conditions of the site, the shaking table was used as gravity separation equipment to treat this refractory gold ore. The efficiency of the shaking table on recovering coarse gold particles was better than other gravity separation equipment, and it had less pollution to the environment. The effects of grinding fineness and number of gravity separations on the index of gravity separation concentrate were investigated. The ball mill process parameters were: steel ball filling rate of 33%, grinding amount of 500 g, and pulp density of 66 wt.%. The process parameters of the shaking table were: 20 kg/h of processing capacity, 20 wt.% of pulp density, 0.65° of lateral gradient, 400 L/h of horizontal flush water, 16 mm of stroke, and 450 min^{-1} of stroke frequency.

The experiment results are shown in Figure 10. Figure 10A indicates that with increase of grinding fineness, the gold grade gradually increased, and the gold recovery to raw ore gradually decreased in the one-stage gravity separation concentrate. After the grinding fineness of −0.074 mm exceeded 73.56%, the gold grade did not rise any more, and the recovery rate of gold did not change much. This also indicated that after the ore reached a certain fineness, the coarser gold particles were substantially dissociated, and the fine particle gold was difficult to achieve monomer dissociation. Therefore, the grinding fineness of −0.074 mm was tentatively determined to be about 73%, and the grinding time was 2 min 15 s. The characteristics of the grinding products, as seen in Table 4, indicated that 88.94% of the gold was distributed below 0. 074 mm.

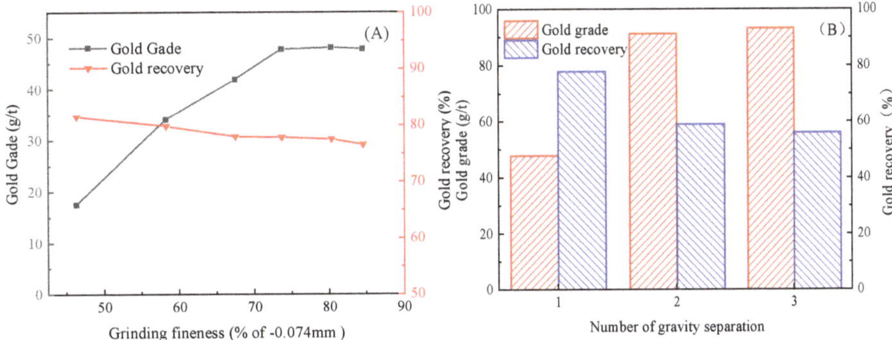

Figure 10. The results of gravity separation as a function of grinding fineness (**A**); and as a function of the number of gravity separations (**B**).

Table 4. The characteristics of the grinding products.

Particle Size (mm)	Yield (%)	Grade (%)				Distribution Rate (%)			
		Au *	As	S	Fe	Au	As	S	Fe
−2.00 + 0.15	1.84	8.67	4.66	6.03	8.11	1.01	1.44	1.45	2.24
−0.15 + 0.106	17.57	5.01	2.41	5.24	4.31	5.55	7.11	12.04	11.37
−0.106 + 0.074	6.87	10.40	4.81	7.94	5.43	4.50	5.55	7.13	5.60
−0.074 + 0.038	31.68	16.83	5.06	6.79	4.82	33.62	26.87	28.10	22.92
−0.038	42.04	20.87	8.37	9.33	9.17	55.32	59.03	51.28	57.87
Total	100.00	15.86	5.96	7.65	6.66	100.00	100.00	100.00	100.00

* The unit of Au is g/t.

The gravity concentrate was subjected to multiple gravity separations after adjusting the height of the table surface. The gold grade of gravity concentrates increased, and the gold recovery decreased with the increasing of number of gravity separations. Since the difference of the index between the two-stage gravity separation and the three-stage gravity separation was not obvious, the number of gravity separations of the shaking table was set at two-stage. The gold grade in the gravity concentrate was about 90 g/t, and the gold recovery rate was about 58%.

3.5. Flotation Separation

The tailings of two stage gravity separation were combined and thickened for flotation feed. The reagent regimes of roughing were investigated, and the results are shown in Figure 11.

The effect of the dosage of sodium silicate on rougher flotation was investigated. The results are presented in Figure 11A. The dosages of other flotation reagents were fixed at 50 g/t of $CuSO_4$, 150 g/t of SBX, and 20 g/t of terpilenol. It can be seen from Figure 11A that with the increase of the dosage of sodium silicate, the gold grade in the flotation coarse concentrate gradually increased, and the recovery rate of gold to the ore gradually decreased. As predicted, a large number of gangues floated during the roughing process, resulting in a decrease in flotation efficiency. The foam was rich and sticky and difficult to combine. After filtration and drying, it could be clearly seen that there were more talc and chlorite in the coarse concentrate. Continuously increasing the dosage of sodium silicate did not effectively depress chlorite and talc. The dosage of sodium silicate was tentatively set at 2000 g/t.

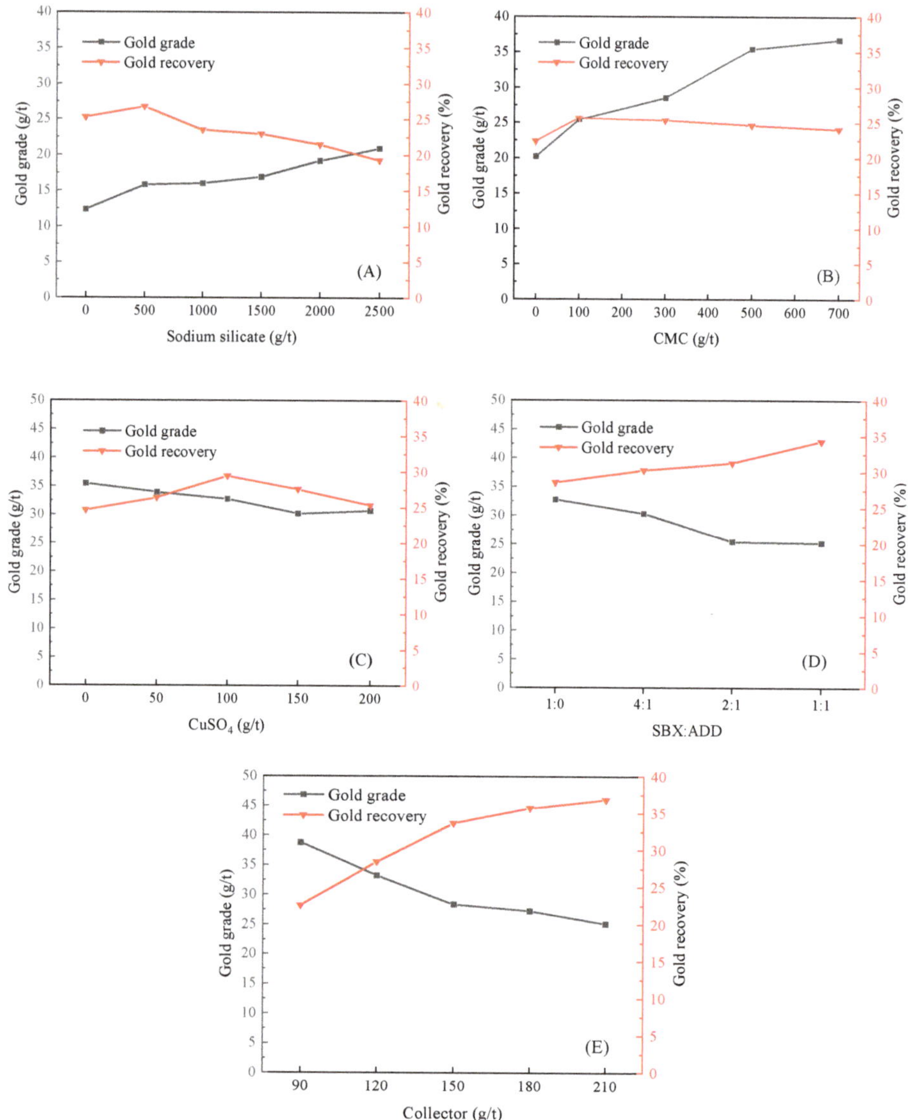

Figure 11. Effects of operating parameters on flotation recovery: (**A**) sodium silicate dosage, (**B**) CMC dosage, (**C**) $CuSO_4$ dosage, (**D**) the ratio of SBX and ADD, (**E**) collector dosage.

CMC has a large molecular weight and a high degree of molecular structure branching and bending. It not only formed a hydrophilic layer on the mineral surface, but also covered up the mineral surface that had formed a hydrophobic layer and a strong inhibiting capacity [40,41]. The addition of CMC enhanced the inhibition of chlorite and talc. The results of the CMC dosage test are shown in Figure 11B. The dosages of other flotation reagents were fixed at 2000 g/t of sodium silicate, 50 g/t of $CuSO_4$, 150 g/t of SBX, and 20 g/t of terpilenol. The results indicated that with the increase of CMC dosage, the gold grade in the coarse concentrate was significantly improved, and the recovery rate of gold decreased less. After the dosage of CMC reached 500 g/t, the gold grade in the coarse concentrate increased slightly, so the dosage of CMC was set to 500 g/t.

Copper sulphate can activate pyrite and arsenopyrite to promote gold recovery in the flotation process. The results of the investigation on the effect of the dosage of $CuSO_4$ on the recovery of gold are shown in Figure 11C. The dosage of other flotation reagents were fixed at 2000 g/t of sodium silicate, 500 g/t CMC, 50 g/t of $CuSO_4$, 150 g/t of SBX, and 20 g/t of terpilenol. As the dosage of $CuSO_4$ was increased, the gold grade decreased slightly, and the recovery rate of gold gradually increased. When the dosage of $CuSO_4$ exceeded 100 g/t, the recovery of gold decreased because the collector was consumed by excess copper ions. The dosage of copper sulfate was set to 100 g/t.

SBX and ADD are often used together to increase the recovery of gold ore. ADD has a certain foaming property and a relatively strong collecting ability. The test results of the mixture ratio of SBX and ADD are shown in Figure 11D. The dosages of flotation reagents were fixed at 2000 g/t of sodium silicate, 500 g/t CMC, 100 g/t of $CuSO_4$, 150 g/t of collector, and 20 g/t of terpilenol. When SBX and ADD were combined as a collector at a ratio of 1:1, the recovery of gold was nearly 6% higher than that of SBX alone, and the gold grade was reduced by nearly 7 g/t. In order to recover gold resources as much as possible, priority was given to the recovery rate of gold, and the ratio of SBX and ADD was set to 1:1.

The test results of the mixed collector dosage are presented in Figure 11E. The dosages of other flotation reagents were fixed at 2000 g/t of sodium silicate, 500 g/t CMC, 100 g/t of $CuSO_4$, and 20 g/t of terpilenol. As the dosage of the collector increased, the recovery rate of gold gradually increased, and the grade of gold gradually decreased. When the dosage of the collector was 210 g/t, the flotation foam was sticky and it was not conducive to cleaner flotation. In general, the optimum dosage of the collector was 180 g/t, and all subsequent experiments were performed with this dosage.

3.6. Closed-Circuit Beneficiation Test

The closed-circuit flow test was carried out on the basis of the condition tests of the gravity separation and flotation. The process is shown in Figure 12, and the test results are shown in Table 5. The results indicated that the gold grade was 91.24 g/t and the gold recovery was 57.58% in the gravity concentrate. The gold grade was 49.44 g/t and the gold recovery was 33.36% in the flotation concentrate, and the total recovery of gold was 90.94%. The main minerals in the flotation concentrate were pyrite and arsenopyrite, the total content of As, S, and Fe reached 88.30%, and the further enrichment space of gold in the gold-bearing mineral was limited by flotation.

Table 5. The results of the closed-circuit test.

Product	Yield (%)	Grade (%)				Recovery (%)			
		Au *	As	Fe	S	Au	As	Fe	S
Gravity concentrate	10.52	91.24	21.31	38.18	35.07	57.58	42.62	38.11	46.00
Flotation concentrate	11.25	49.44	18.36	36.36	33.58	33.36	39.27	38.81	47.10
Tailing	78.23	1.93	1.22	3.11	0.71	9.06	18.11	23.08	6.89
Feed	100.00	16.67	5.26	10.54	8.02	100.00	100.00	100.00	100.00

* The unit of Au is g/t.

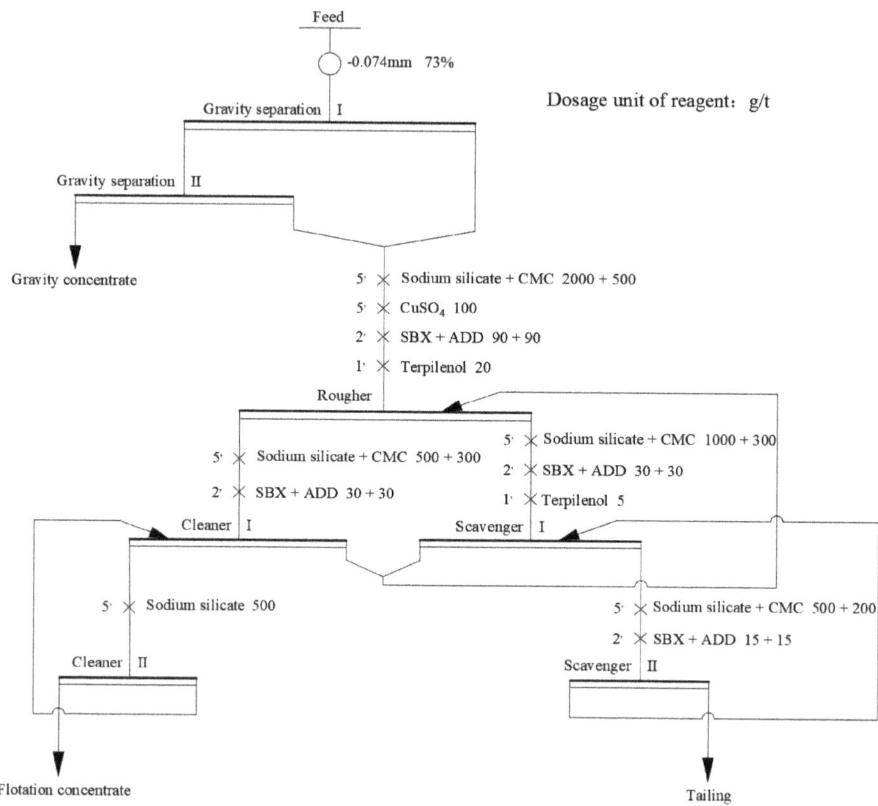

Figure 12. Image of closed-circuit combination process of gravity separation and flotation.

4. Conclusions

(A) The results of the processing mineralogy study showed that the gold grade in the ore was about 15.96 g/t, and the gold mainly existed in the form of natural gold and electrum. The main gold-bearing materials were pyrite, arsenopyrite, and some gangues. The gold particles were uneven, and most of them were fine grains of 5–10 μm. Other metallic minerals in the ore included sphalerite, galena, and jamesonite, and gangue minerals included quartz, mica, feldspar, talc, and chlorite.

(B) Fine-grained gold, which was highly dispersed in the ore-bearing minerals, and the large amount of stratified silicate minerals were the main reasons for the poor pretreatment effect of the refractory gold ore whether it was gravity separation alone or flotation alone. Fine-grained gold in arsenopyrite is difficult to dissociate, and the removal of arsenic during the beneficiation process will lead to a decrease in gold recovery

(C) According to the research results of the processing mineralogy study, the gravity–flotation combined beneficiation process was used to pretreat the refractory gold deposit, and the coarse gold particles and fine gold particles in sulphides were recovered, respectively. The index of the closed-circuit test of the gravity–flotation combined beneficiation was that the gold grade was 91.24 g/t and the gold recovery was 57.58% in the gravity concentrate, the gold grade was 49.44 g/t and the gold recovery was 33.36% in the flotation concentrate, and the total recovery of gold was 90.94%. This indicated that gravity–flotation combined beneficiation was an effective pretreatment method for this refractory gold ore. The effect of the gravity–flotation combined process was superior to the gravity beneficiation process and the flotation process, and the cost was lower than the full-process flotation.

(D) Limited by the occurrence of the refractory gold deposit, the gold concentrate obtained is difficult to be further enriched through the method of beneficiation. The gold concentrate obtained by this process will eventually be shipped back for metallurgical processing to obtain gold.

Author Contributions: X.W. and W.Q. conceived and designed the experiments; X.W., F.J., C.Y., W.L., Z.Z., and H.S. performed the experiments; X.W. and Y.C. analyzed the data; X.W. and C.Y. wrote the paper.

Funding: This research was funded by Provincial Science and technology leader (Innovation team of interface chemistry of efficient and clean utilization of complex mineral resources, Grant No. 2016RS2016); National Natural Science Foundation of China (Project No. 51604302 and No. 51574282); and the Key laboratory of Hunan Province for Clean and Efficiency Utilization of strategic Calcium-containing mineral Resources (No. 2018TP1002).

Acknowledgments: We would like to thank the Central South University for providing us with the experimental platform.

Conflicts of Interest: The authors declare no conflict of interest.

References

1. O'Connor, C.T.; Dunne, R.C. The flotation of gold bearing ores—A review. *Miner. Eng.* **1994**, *7*, 839–849. [CrossRef]
2. Li, W.J.; Song, Y.S.; Chen, Y.; Cai, L.L.; Zhou, G.Y. Beneficiation and leaching study of a muti-Au carrier and low grade refractory gold ore. *IOP Conf. Series Mater. Sci. Eng.* **2017**, *231*, 12169. [CrossRef]
3. Guo, X.; Xin, Y.; Wang, H.; Tian, Q. Mineralogical characterization and pretreatment for antimony extraction by ozone of antimony-bearing refractory gold concentrates. *Trans. Nonferrous Metals Soc. China* **2017**, *27*, 1888–1895. [CrossRef]
4. Allan, B.; Woodcock, J.T. A review of the flotation of native gold and electrum. *Miner. Eng.* **2001**, *9*, 931–962. [CrossRef]
5. Asamoah, R.K.; Skinner, W.; Addai-Mensah, J. Leaching behaviour of mechano-chemically activated bio-oxidised refractory flotation gold concentrates. *Powder Technol.* **2018**, *331*, 258–269. [CrossRef]
6. Stewart, M.; Kappes, D. SART for copper control in cyanide heap leaching. *J. South. Afr. Inst. Min. Metall.* **2012**, *112*, 1037–1043.
7. Oraby, E.A.; Eksteen, J.J. Gold dissolution and copper suppression during leaching of copper–gold gravity concentrates in caustic soda-low free cyanide solutions. *Miner. Eng.* **2016**, *87*, 10–17. [CrossRef]
8. Drif, B.; Taha, Y.; Hakkou, R.; Benzaazoua, M. Recovery of Residual Silver-Bearing Minerals from Low-Grade Tailings by Froth Flotation: The Case of Zgounder Mine, Morocco. *Minerals* **2018**, *8*, 273. [CrossRef]
9. Guo, B.; Peng, Y.; Espinosa-Gomez, R. Effects of free cyanide and cuprous cyanide on the flotation of gold and silver bearing pyrite. *Miner. Eng.* **2015**, *71*, 194–204. [CrossRef]
10. Xie, X.; Fu, Y.; Liu, J. Mineralogical characteristic of ductile-brittle shear zone in the Baguamiao gold deposit, Shaanxi Province: Implication for gold mineralization. *Geol. J.* **2017**, *52*, 81–96. [CrossRef]
11. Reyes, N.R.R.; Echeverry-Vargas, L.; Cataño-Martínez, J. Characterization by QEMSCAN and FE-SEM of ore bodies gold artisanally treated with mercury in Antioquia. Colombia. *Rev. Prospect.* **2017**, *15*, 107–116. [CrossRef]
12. Rabieh, A.; Albijanic, B.; Eksteen, J.J. A review of the effects of grinding media and chemical conditions on the flotation of pyrite in refractory gold operations. *Miner. Eng.* **2016**, *94*, 21–28. [CrossRef]
13. Parbhakar-Fox, A. Geoenvironmental Characterisation of Heap Leach Materials at Abandoned Mines: Croydon Au-Mines, QLD, Australia. *Minerals* **2016**, *6*, 52. [CrossRef]
14. Yalcin, E.; Kelebek, S. Flotation kinetics of a pyritic gold ore. *Int. J. Miner. Process.* **2011**, *98*, 48–54. [CrossRef]
15. Dominy, S. Geometallurgical Study of a Gravity Recoverable Gold Orebody. *Minerals* **2018**, *8*, 186. [CrossRef]
16. McGrath, T.D.H.; Connor, L.O.; Eksteen, J.J. A comparison of 2D and 3D shape characterisations of free gold particles in gravity and flash flotation concentrates. *Miner. Eng.* **2015**, *82*, 45–53. [CrossRef]
17. Ofori-Sarpong, G.; Amankwah, R.K. Comminution environment and gold particle morphology: Effects on gravity concentration. *Miner. Eng.* **2011**, *24*, 590–592. [CrossRef]
18. Jena, M.S.; Mohanty, J.K.; Sahu, P.; Venugopal, R.; Mandre, N.R. Characterization and Pre-concentration of Low Grade PGE Ores of Boula Area, Odisha using Gravity Concentration Methods. *Trans. Indian Inst. Met.* **2017**, *70*, 287–302. [CrossRef]

19. Koppalkar, S.; Bouajila, A.; Gagnon, C.; Noel, G. Understanding the discrepancy between prediction and plant GRG recovery for improving the gold gravity performance. *Miner. Eng.* **2011**, *24*, 559–564. [CrossRef]

20. Kononova, O.N.; Kholmogorov, A.G.; Kononov, Y.S.; Pashkov, G.L.; Kachin, S.V.; Zotova, S.V. Sorption recovery of gold from thiosulphate solutions after leaching of products of chemical preparation of hard concentrates. *Hydrometallurgy* **2001**, *59*, 115–123. [CrossRef]

21. Rabieh, A.; Albijanic, B.; Eksteen, J. Influence of grinding media and water quality on flotation performance of gold bearing pyrite. *Miner. Eng.* **2017**, *112*, 68–76. [CrossRef]

22. Yang, X.; Albijanic, B.; Liu, G.; Zhou, Y. Structure–activity relationship of xanthates with different hydrophobic groups in the flotation of pyrite. *Miner. Eng.* **2018**, *125*, 155–164. [CrossRef]

23. Ejtemaei, M.; Nguyen, A.V. Characterisation of sphalerite and pyrite surfaces activated by copper sulphate. *Miner. Eng.* **2017**, *100*, 223–232. [CrossRef]

24. Chen, J.; Chen, Y.; Wei, Z.; Liu, F. Bulk flotation of auriferous pyrite and arsenopyrite by using tertiary dodecyl mercaptan as collector in weak alkaline pulp. *Miner. Eng.* **2010**, *23*, 1070–1072. [CrossRef]

25. Badri, R.; Zamankhan, P. Sulphidic refractory gold ore pre-treatment by selective and bulk flotation methods. *Adv. Powder Technol.* **2013**, *24*, 512–519. [CrossRef]

26. Tapley, B.; Yan, D. The selective flotation of arsenopyrite from pyrite. *Miner. Eng.* **2003**, *16*, 1217–1220. [CrossRef]

27. Li, Y.; Liu, Z.; Li, Q.; Liu, F.; Liu, Z. Alkaline oxidative pressure leaching of arsenic and antimony bearing dusts. *Hydrometallurgy* **2016**, *166*, 41–47. [CrossRef]

28. Gu, K. Arsenic and antimony extraction from high arsenic smelter ash with alkaline pressure oxidative leaching followed by Na2S leaching. *Sep. Purif. Technol.* **2019**, *222*, 53–59. [CrossRef]

29. Gu, K.; Li, W.; Han, J.; Liu, W.; Qin, W.; Cai, L. Arsenic removal from lead-zinc smelter ash by NaOH-H_2O_2 leaching. *Sep. Purif. Technol.* **2019**, *209*, 128–135. [CrossRef]

30. Marsden, J.; House, I. *The Chemistry of Gold Extraction*; Ellis Horwood: Maylands Avenue, UK, 2006.

31. Raphulu, M.C.; Scurrell, M.S. Cyanide leaching of gold catalysts. *Catal. Commun.* **2015**, *67*, 87–89. [CrossRef]

32. Oraby, E.; Eksteen, J.; Tanda, B. Gold and copper leaching from gold-copper ores and concentrates using a synergistic lixiviant mixture of glycine and cyanide. *Hydrometallurgy* **2017**, *169*, 339–345. [CrossRef]

33. Murthy, D.; Kumar, V.; Rao, K. Extraction of gold from an Indian low-grade refractory gold ore through physical beneficiation and thiourea leaching. *Hydrometallurgy* **2003**, *68*, 125–130. [CrossRef]

34. Asamoah, R.K.; Skinner, W.; Addai-Mensah, J. Alkaline cyanide leaching of refractory gold flotation concentrates and bio-oxidised products: The effect of process variables. *Hydrometallurgy* **2018**, *179*, 79–93. [CrossRef]

35. Fomchenko, N.V.; Kondrat'Eva, T.F.; Muravyov, M.I.; Kondrat'Eva, T.F. A new concept of the biohydrometallurgical technology for gold recovery from refractory sulfide concentrates. *Hydrometallurgy* **2016**, *164*, 78–82. [CrossRef]

36. Mubarok, M.; Winarko, R.; Chaerun, S.; Rizki, I.; Ichlas, Z. Improving gold recovery from refractory gold ores through biooxidation using iron-sulfur-oxidizing/sulfur-oxidizing mixotrophic bacteria. *Hydrometallurgy* **2017**, *168*, 69–75. [CrossRef]

37. Yang, H.-Y.; Liu, Q.; Chen, G.-B.; Tong, L.-L.; Ali, A. Bio-dissolution of pyrite by Phanerochaete chrysosporium. *Trans. Nonferrous Met. Soc. China* **2018**, *28*, 766–774. [CrossRef]

38. Ciftci, H.; Akcil, A. Biohydrometallurgy in Turkish gold mining: First shake flask and bioreactor studies. *Miner. Eng.* **2013**, *46*, 25–33. [CrossRef]

39. Wei, G.; Bo, F.; Jinxiu, P.; Wenpu, Z.; Xianwen, Z. Depressant behavior of tragacanth gum and its role in the flotation separation of chalcopyrite from talc. *J. Mater. Res. Technol.* **2019**, *8*, 697–702. [CrossRef]

40. Feng, B.; Feng, Q.; Lu, Y.; Gu, Y. The effect of PAX/CMC addition order on chlorite/pyrite separation. *Miner. Eng.* **2013**, *42*, 9–12. [CrossRef]

41. Feng, Q.-M.; Feng, B.; Lu, Y.-P. Influence of copper ions and calcium ions on adsorption of CMC on chlorite. *Trans. Nonferrous Met. Soc. China* **2013**, *23*, 237–242. [CrossRef]

MDPI

St. Alban-Anlage 66

4052 Basel

Switzerland

Tel. +41 61 683 77 34

Fax +41 61 302 89 18

www.mdpi.com

Minerals Editorial Office

E-mail: minerals@mdpi.com

www.mdpi.com/journal/minerals